Chronology of KSC and KSC Related Events for 2000

Elaine E. Liston, Donna A. Atkins, Deborah A. Guelzow,
InDyne, Inc., Kennedy Space Center, Florida

NASA/TM-2000-208590   February 2001

FOREWORD

This 2000 Chronology is published to describe and document Kennedy Space Center's role in NASA's progress.

Materials for this Chronology were selected from a number of published sources. The document records KSC events of interest to historians and other researchers. Arrangement is by date of occurrence, though the source cited may be dated one or more days after the event.

Materials were researched and prepared for publication by Archivist Elaine E. Liston, Librarians Donna A. Atkins and Deborah A. Guelzow. Production assistance was provided by Barbara E. Green, Library Technician.

Comment on the Chronology should be directed to the John F. Kennedy Space Center, Archives, LIBRARY-E, Kennedy Space Center, Florida 32899. The Archivist may also be reached by e-mail at Elaine.Liston-1@ksc.nasa.gov or by phone at (321) 867-2407.

TABLE OF CONTENTS

January 1

February 6

March 10

April 16

May 21

June 27

July 32

August 38

September.. 44

October 48

November 65

December 74

JANUARY

JANUARY 1: NASA continues to be "green," meaning Agency systems have not been substantively affected by any problems during the year-end transition. During the primary monitoring period (8 a.m. EST Dec. 31 through 3 a.m. EST Jan. 1) the Agency suffered a few minor anomalies that were easily fixed. Only one, involving a piece of planning software, appeared to be Y2K related, and it did not affect any mission-critical systems. Flight controllers continue to make contact with NASA spacecraft according to previously planned schedules. Remaining NASA spacecraft, which have been configured so as not to require commanding over the Y2K transition, will be contacted by controllers over the next several days. The Johnson Space Center reports that the Mission Control Center for the International Space Station, which was taken offline before the Y2K transition in Moscow (4 p.m. EST Dec. 31), was successfully brought back online. Also over the next few days, NASA will continue to monitor its infrastructure and business systems. The Agency expects to resume business as usual on Monday, January 3. ["NASA Y2K Status Report for Midnight EST," *NASA Y2K Status Report for Midnight EST*, January 1, 2000.]

◆ A New Year's Eve party, called "Blast Into the Millennium", was held at Cape Canaveral Air Station, Launch Complex 14. The whole idea was to stage a special salute to the elite fraternity of aerospace workers who made America's earliest forays in to the final frontier a reality. "We were looking for the most historic place we could find to bring in the new millennium and celebrate our achievements in space, and I cannot think of a more historic site," said NBC News correspondent and party organizer Jay Barbree. [Todd Halvorson. (2000). Space Luminaries Ring in the New Year at Historic Launch Pad [Online]. Available WWW: http://www.space.com/ [2000, December 30].]

JANUARY 3: Warren I. Wiley has been named Acting Director of Engineering Development. [Memoranda (NASA). Roy D. Bridges, Jr. Subject: "Key Personnel Changes," January 3, 2000.]

JANUARY 4: Florida officials have offered Beal Aerospace Technologies Inc., a package of tax incentives and grants valued at about $10 million to get it to build a manufacturing and launch facility at Cape Canaveral. ["Florida tries to lure rocket maker," *Florida Today*, January 5, 2000, p 12C.]

JANUARY 5: NASA may fly an additional shuttle mission to the International Space Station this spring to perform maintenance work. ["NASA may launch additional shuttle," *Florida Today*, January 6, 2000, p 7A.]

JANUARY 6: Processing of Space Shuttle Endeavour for mission STS-99 has resumed at Launch Pad 39A following the holiday downtime. ["Endeavour launch team resumes launch preparations," *KSC Countdown*, January 6, 2000.]

JANUARY 11: The STS-99 Shuttle Radar Topography Mission (SRTM) crew arrives today for Terminal Countdown Demonstration Test activities. The six-member crew comprises Commander Kevin Kregel, Pilot Dominic Gorie, and Mission Specialists Janet Kavandi, Janice Voss, Mamoru Mohri, and Gerhard Thiele. Mohri is with the National Space

Development Agency (NASDA) of Japan and Thiele is with the European Space Agency. ["STS-99 crew at KSC for TCDT activities," *KSC Countdown*, January 11, 2000.]

◆ Don McMonagle, Manager, Launch Integration, Space Shuttle Program Office, will leave NASA on Jan. 14, 2000. ["Did You Know?", *KSC Countdown*, January 11, 2000.]

◆ In a major setback for the International Space Station, the launch of a key Russian module will be delayed until late summer because flaws have been detected in a rocket that is to carry it into orbit. ["Rocket flaws delay station," *Florida Today*, January 12, 2000, p 1A.]

JANUARY 12: Kevin Kregel, commander of NASA's next shuttle mission jokingly told German astronaut Gerhard Thiele that "Rommel would have been proud of you" after Thiele drove an armored personnel carrier during training Wednesday. Kregel made the remark about the German military leader on a vacant launch pad at Kennedy Space Center where crew members were practicing shuttle escape maneuvers. Part of their training includes each crew member driving an M-113 armored personnel carrier away from the launch pad.["German astronaut gains praise in drill," *Florida Today*, January 13, 2000, p 6A.]

◆ The U. S. government is suing three companies it claims concealed fraud by a subcontractor that spent millions in space shuttle program money on homes, jewelry and exotic vacations. The Justice Department suit, filed Tuesday, named Rockwell International, Inc., Boeing North American Inc. and United Space Alliance, the current shuttle contractor. ["U.S. alleges fraud, sues 3 shuttle contractors," *Florida Today*, January 13, 2000, p 1A.]

◆ Concerned that Russia won't be able to launch its next piece of the International Space Station this year, U. S. Rep. Dave Weldon is calling for NASA to fly a U.S.-built replacement segment instead. The backup component, called the Interim Control Module, could be launched aboard a shuttle late this year to provide propulsion for the station, Weldon said Wednesday. ["NASA should fly U.S. -built module instead," *Florida Today*, January 13, 2000, p 1A.]

◆ The U.S. Postal Service unveiled its 1980s "Celebrate the Century" stamps, including one honoring the nation's space program, Wednesday at the Kennedy Space Center Visitor Complex. The test on the gummed side of the space Shuttle stamp reads, "Space Shuttles have transformed U.S. space exploration. These reusable crafts can launch satellites and house labs for scientific experiments. Columbia, the first Space Shuttle, was launched April 12, 1981." ["Space Shuttle Program Joins American Icons In Commemorative Stamp Collection," *NASA News Release #00-12*, January 12, 2000.]

JANUARY 13: Today the STS-99 crew will practice emergency egress procedures at Launch Pad 39A. ["TCDT to end with test countdown for crew and Shuttle engineers," *KSC Countdown*, January 13, 2000.]

◆ Center Director Roy Bridges will moderate a discussion on the future of space as it relates to the state of Florida. Kennedy Space Center's Visitor's Complex will be the stage for the first Florida Space Summit on Friday, Jan. 14 from noon to 2:30 p.m. Participants include Senator Bob Graham, Rep. Dave Weldon, members of Florida's State government including Gov. Jeb Bush, NASA Administrator Dan Goldin, KSC Director Roy Bridges,

45th Space Wing Commander Brig. Gen. Donald Pettit and heads of aerospace companies. ["Kennedy Space Center to host first Florida Space Summit," *NASA News Release #5-00*, January 13, 2000.]

◆ NASA on Thursday secured a launch date for shuttle Endeavour, clearing the way for a Jan. 31 flight to map the Earth in unmatched detail. To make it possible, Air Force officials agreed to postpone a temporary shutdown of the military tracking system that monitors all shuttle launches and landings. ["Shuttle launch date secured," *Florida Today*, January 14, 2000, p 10A.]

◆ Hughes Electronics Corp. is getting out of the aerospace industry to focus on entertainment and Internet services, striking a $3.75 billion deal with Boeing Co. on Thursday to sell its satellite-building business. The Boeing Co. has a major aerospace division at the Kennedy Space Center, where it employees about 2,500 workers. It wasn't immediately known if the Hughes acquisition had any local ramifications. ["Boeing to buy Hughes satellite business," *Florida Today*, January 14, 2000, p _____.]

JANUARY 14: The Florida Space Summit brought together political and space power brokers to discuss how to make the state a flourishing launch site during the 21st century. The 2 ½-hour gathering at the KSC Visitor Complex produced no revelations about the challenges Florida faces. But participants said the teamwork forged at the summit will boost efforts to make Florida the world's dominant launch site. "This conference is unprecedented," NASA Administrator Dan Goldin said. ["Politicians, space leaders aim to lure launch business," *Florida Today*, January 15, 2000, p 1A & 4A.]

◆ Shuttle Endeavour astronauts practiced a launch countdown at launch pad 39A. Endeavour is scheduled to lift off Jan. 31. ["Crew prepares for Jan. 31 liftoff," *Florida Today*, January 15, 2000, p 1A.]

JANUARY 17: Edward Sheffield, 71, died Monday. While employed in the aerospace industry, Sheffield worked as a department manager for Pan Am at Cape Canaveral Air Station and Federal Electric Corp., RCA Service Co., Lockheed Space Operations Co., and Lockheed Martin, all at Kennedy Space Center. ["Retired NASA manager Edward Sheffield dies," *Florida Today*, January 20, 2000, p 3B.]

◆ A pair of new transporters for Shuttle payload canisters arrived by barge on Monday, Jan. 17 from their manufacturer, the KAMAG Transporttechnik, GmbH, of Ulm, Germany. ["New Payload Transporters arrive at KSC," *NASA News Release #9-00*, January 21, 2000.]

JANUARY 18: During a flight readiness review on Tuesday, shuttle officials made Endeavour's launch contingent on finishing their investigation of protective heat tiles before firming up the spaceship's planned Jan. 31 launch date. One tile fell off the wing of sister ship Discovery during its landing at Kennedy Space Center Dec. 27. ["NASA checking shuttle's heat tiles," *The Orlando Sentinel*, January 19, 2000, p A-10.]

JANUARY 20: The USAF, supported by numerous contractors and private agencies, continues the year-long celebration of the first launch from Florida: Bumper 8, a modified German V-2 rocket, launched on July 24, 1950. The next monthly celebration is the dedication by the U.S. Space Walk of Fame in Titusville of a plaque. ["Celebration of the First Launch From Florida," *KSC Countdown*, January 20, 2000.]

◆ A $200 million Defense Satellite Communications System spacecraft was carried aloft Thursday night on a Lockheed Martin Atlas rocket launched from Cape Canaveral Air Station. Gusty winds died down in time for the 8:03 p.m. liftoff. It was the first launch from the Cape in 2000, but another 19 rocket missions are currently scheduled to take off. ["Military launches satellite to join defense network," *Florida Today*, January 21, 2000, p 4A.]

JANUARY 21: The "NewsCapade with Al Neuharth" exhibit opens today at the Kennedy Space Center Visitor Complex. Housed in a 2,000 square-foot pavillion, NewsCapade is a traveling version of the Newseum in Arlington, VA. ["KSC welcomes NewsCapade," *Florida Today*, January 21, 2000, p 1B.]

◆ NASA has cut one day of science work from shuttle Endeavour's upcoming mission as a precaution against problems with the 200-foot mast that will be used to make a detailed radar map of the Earth. Now set for launch Jan. 31, Endeavour still will fly in space for 11 days. But the crew will retract the radar mast one day earlier than originally planned, officials said Friday. ["NASA chops work day from shuttle mission," *Florida Today*, January 22, 2000, p 3A.]

JANUARY 24: Shuttle Endeavour is fit for its Jan. 31 launch, NASA officials said Monday after reviewing inspections of the spaceship's tiles. Two tiles have been replaced, but officials said the others appear to be in good condition. Launch is set between 12:47 and 2:49 p.m. Monday from KSC. ["Endeavour fit to fly Monday, NASA declares," *Florida Today*, January 25, 2000, p 7A.]

JANUARY 27: Cold weather this week is not expected to affect Endeavour. The orbiter's reaction control system heaters are remaining powered, which is standard cold weather configuration. New cameras have been installed on the forward and aft bulkheads and the payload bay doors were expected to be closed yesterday. ["Endeavour expected to weather cold; on track for launch Jan. 31," *KSC Countdown*, January 27, 2000.]

◆ NASA may send a shuttle to the International Space Station in April instead of waiting for Russia to launch the outpost's next segment first, officials said Thursday. A crew on shuttle Atlantis would fly the proposed mission, which calls for routine maintenance to prepare the station for the next addition – the much-delayed Russian Service Module. ["NASA may fly shuttle to space station in April," *Florida Today*, January 28, 2000, p 1A.]

◆ Shuttle Endeavour's astronauts arrived at Kennedy Space Center Thursday, ready for an 11-day voyage to map the world's peaks and valleys better than ever before. A three-day countdown is to begin tomorrow for the 12:47 p.m. launch Monday. ["Crew arrives at space center; shuttle liftoff set for Monday." *Florida Today*, January 28, 2000, p 5A.]

JANUARY 28: Hosted by Brevard County and the KSC Support Committee, the first Florida Space summit marks the first time Central Florida counties have gathered to forge a unified front to push for more funding for Kennedy Space Center and Cape Canaveral. Commissioners from Indian River, Lake, Osceola, Seminole and Volusia counties also came to the afternoon-long retreat, which included a tour of KSC. ["Summit pushes for more space funding," _Florida Today_, January 29, 2000, p 1B.]

◆   The countdown clock started ticking Friday toward the launch of shuttle Endeavour, but an engine problem during a December mission could keep the flight temporarily grounded. Endeavour is scheduled to lift off from Kennedy Space Center on Monday on an 11-day radar mapping mission of Earth. Before the shuttle gets the green light to launch, however, mission managers want to clear up concerns about of the three main engines that powered sister ship Discovery last month. ["Engine trouble makes launch of shuttle iffy," _The Orlando Sentinel_, January 29, 2000, p A-20.]

◆   A ceremony at the Astronaut Memorial Plaza at Sand Point Park, Titusville, honored the seven astronauts who died in the space shuttle Challenger accident on Jan. 28, 1986, and the three astronauts who died in the Apollo 1 accident on Jan. 27, 1967. ["Ceremony honors astronauts," _Florida Today_, January 29, 2000, p 1B & 2B.]

JANUARY 30: NASA officials cleared shuttle Endeavour for flight Sunday after wrapping up their analysis of how a defective part wound up inside one of Discovery's engines last month. While the problem appears resolved, the weather may pose another problem. A cold front passing through Florida has moved much more slowly than expected. As a result, the chance of favorable conditions is only 40 percent. ["Shuttle gets go-ahead," _The Orlando Sentinel_, January 31, 2000, p A-4.]

JANUARY 31: Rain and thick clouds scrubbed shuttle Endeavour's planned liftoff Monday. Also, during the final stages of Monday's countdown, flight controllers detected a problem with one of Endeavour's twin master events controllers, a pair of small redundant electronics boxes in the orbiter's tail section. ["Weather, computer part scrub launch," _The Orlando Sentinel_, February 1, 2000, p A-7.]

◆   Among the VIPs at KSC for launch of STS-99 were Jonathan Pierce and Ralph Charles. Pierce is a 10-year-old boy who wears a protective cooling suit designed by NASA. Charles is a 100-year-old pilot who has experienced the history of flight from its origins on the dunes of North Carolina to travels in space. He knew Orville Wright and worked with Charles Lindbergh. ["10-year old and 100-year old among VIPs at KSC for launch," _KSC Countdown_, February 1, 2000.]

DURING JANUARY: Space shuttle launch delays due to wiring problems last summer as well as schedule problems with some mission hardware have caused perhaps the largest buildup of space payloads at the Kennedy Space Center in the 35-year history of the facility. About 10 major payloads are in checkout at Kennedy – nearly two years worth of launch operations. ["Payload buildup," _Aviation Week & Space Technology_, January 24, 2000, p 19.]

FEBRUARY

FEBRUARY 1: NASA managers decided Tuesday to replace a critical computer relay in shuttle Endeavour before launching the STS-99 radar mapping mission of Earth. Liftoff has been rescheduled for no earlier than Feb. 9. ["Shuttle struggles for launch," *The Orlando Sentinel*, February 2, 2000, p A-8.]

FEBRUARY 2: NASA officials have rescheduled the launch of shuttle Endeavour for no earlier than Feb. 11. Although the 11-day radar-mapping mission should be ready to lift off by then, a bottleneck on the Air Force's Eastern Range likely will force Endeavour to wait until at least Feb. 12, during a launch window from 12:25 to 2:39 p.m. ["Endeavour grounded until mid-February," *The Orlando Sentinel*, February 3, 2000, p A-11.]

◆ The STS-99 astronauts returned to Houston. Commander Kregel and Pilot Gorie departed from KSC Shuttle Landing Facility in T-38 jets, and the rest of the crew left from Patrick Air Force Base. The flight crew will return to KSC four days prior to launch. ["Shuttle Update," *KSC Countdown*, February 3, 2000.]

FEBRUARY 3: A Lockheed Martin Atlas rocket carried a $200 million television satellite into space at 6:30 p.m. Thursday from Cape Canaveral Air Station. ["Atlas carries TV satellite," *Florida Today*, February 4, 2000, p 11A.]

◆ NASA will launch a U.S. replacement module to the International Space Station in December if Russia does not deliver a critical piece to the outpost this summer, agency Administrator Dan Goldin said Thursday. Goldin said if Russia has not launched the Service Module this summer that NASA would do the final work on its replacement module and send it aloft on a shuttle in December. ["NASA may send U.S. module to station," *Florida Today*, February 4, 2000, p 1A.]

FEBRUARY 4: First the first time since 1994, Kennedy Space Center will increase its government work force by hiring 158 engineers and technicians at the center, KSC Director Roy Bridges said Friday. The new employees – to be hired by Sept. 30 – are intended to bolster a KSC work force that has lost many experienced workers in recent years because of budget cuts. Bridges said the new positions are being added, in part, to protect the safety of the space shuttle fleet. ["KSC to hire 158 new workers," *Florida Today*, February 5, 2000, p 1A.]

◆ The word "force" was returned to the Cape Canaveral Air Force Station's title on Friday, reversing the military's 1992 decision to drop the word. ["It's now Cape Canaveral Air Force Station," *Florida Today*, February 5, 2000, p 8E.]

FEBRUARY 7: STS-99 crew returned to KSC to prepare for launch on Friday, Feb. 11. Launch Countdown begins at 5:30 p.m. tomorrow, Feb. 8. ["Crew arrives for STS-99 launch - countdown begins 5:30 p.m. today," *KSC Countdown*, February 8, 2000.]

◆ Kennedy Space Center employees received $140,000 in Space Act Awards for the 1999 fiscal year, an increase of about $43,000 over 1998. KSC led all Centers with 65 software releases, a new Agency record. ["NASA technology awards continued 7-year trend of increases," *KSC Countdown*, February 8, 2000.]

◆   NASA would see its budget increase for the first time in seven years and have enough money to hire new workers if Congress agrees to the $14 billion spending request President Clinton unveiled Monday. The request represents a 3 percent increase, or $435 million more than spending in the current fiscal year. In announcing the spending request in a briefing at NASA headquarters, Administrator Dan Goldin said work force downsizing at key human space flight centers, including Kennedy Space Center, has stopped for now. ["Plan offers 3 percent rise in NASA funding," *Florida Today*, February 8, 2000, p 1A & 2A.]

◆   A new granite wall has been added to the Astronauts Memorial Space Mirror at the KSC Visitor Complex. The six-foot by six-foot wall features laser-engraved photos and biographies of the 17 fallen astronauts honored in the Space Mirror. The mirror, 42.5-feet wide and 50-feet tall, was unveiled in May 1991. The granite wall was completed in December 1999 and unveiled for the public Feb. 7, 2000. ["Wall of Space History," *KSC Countdown*, February 22, 2000.]

FEBRUARY 8: Four cellular telephone satellites were launched into orbit aboard a Boeing Delta II rocket from Cape Canaveral Air Force Station on Tuesday. The $110 million mission was the first Delta launch of the year from the Cape and the fifth successful Delta Globalstar launch since June. ["Delta II launch completes Globalstar satellite fleet," *The Orlando Sentinel*, February 9, 2000, p A-4.]

FEBRUARY 10: NASA security evacuated the Headquarters Building at Kennedy Space Center on Thursday after callers reported seeing two suspicious-looking people in the building. It turned out to be a false alarm, and the people in question were properly badged NASA employees dressed in hooded coats and wearing backpacks, KSC spokesman Bruce Buckingham said. The workers identified themselves while waiting outside in a parking lot as security guards searched the building room by room. Officials also began stopping all cars leaving the space center and checking for employee identification. ["KSC man office evacuated," *Florida Today*, February 11, 2000, p 1B.]

◆   NASA's race to become "faster, better and cheaper" has left its human spaceflight team too small and too inexperienced to safely cope with the high rate of shuttle flights that will be needed to build the International Space Station, the independent Aerospace Safety Advisory Panel reported Thursday. "We have found continuing work force problems at Kennedy Space Center, Johnson Space Center and Marshall Space Flight Center related to Space Shuttle operations and the launching of the International Space Station," the report authors wrote. The panel also faulted NASA and United Space Alliance for relaying solely on anticipated productivity improvements to meet increasing launch demands. ["Inexperienced, small work force concerns panel," *Florida Today*, February 11, 2000, p 10A.]

◆   More than two dozen Boeing Co. workers picketed Thursday at Cape Canaveral Air Force Station joining their union colleagues more than 3,000 miles away in a strike against the Seattle-based aerospace giant. The workers – a group of engineers and technicians for Boeing's aerospace services unit at the Cape – protested what they described as the company's intolerable attempt to wipe out valuable benefits and salary gains as part of a new contract. ["Boeing workers picket at Cape," *The Orlando Sentinel*, February 11, 2000, p B-1 & B-5.]

FEBRUARY 11: Shuttle Endeavour's six astronauts blasted into orbit through sunny, blue skies above Kennedy Space Center to begin their long-awaited 11-day radar-mapping mission of Earth. The launch of STS-99 began from Launch Complex 39A at 12:43 p.m. EST. ["A look back at Earth," *The Orlando Sentinel*, February 12, 2000, p A-1 & A-10.]

◆ A week after NASA Administrator Dan Goldin threatened to proceed without them, Russian government officials Friday promised to complete their key International Space Station component by early July. The Russians' vow came after a two-day regularly scheduled program review in Moscow between Russian and U.S. officials. ["Russians vow to complete key module by early July," *Florida Today*, February 12, 2000, p 4A.]

FEBRUARY 14: "Race to Space," is a motion picture currently being filmed at Kennedy Space Center. ["'Race to Space' actor entertains at Mango Tree," *Florida Today*, February 15, 2000, p 1D.]

FEBRUARY 19: The Kennedy Space Center Visitor Complex opened the Dr. Kurt H. Debus Conference Center and three new exhibits – Astronaut Encounter, Exploration in the New Millennium and Earth Space Exploration. They are the final pieces of a $120 million redevelopment project. ["KSC exhibits offer hands-on experience," *Florida Today*, February 21, 2000, p 1B & 2B.]

FEBRUARY 21: NASA has set dates and named crews for two planned shuttle Atlantis flights to the International Space Station in the spring and summer. The first Atlantis mission (STS-101) is scheduled no earlier than April 13, and the second (STS-106) no earlier than Aug. 19. Other than the two Atlantis missions and a planned Endeavour flight of STS-97 in November to carry an Italian-built logistics module to the station, the remainder of the shuttle schedule through 2000 remains to be firmed up. ["NASA sets two shuttle flights," *Florida Today*, February 22, 2000, p 2A.]

FEBRUARY 22: The next scheduled mission, STS-101, is targeted no earlier than April 13. The seven-member crew will be preparing the Space Station for the arrival of the Zvezda service module, expected to be launched by Russia in July. NASA has confirmed plans to fly an additional mission to the Station this year, STS-106, in August to transfer supplies and outfit the Station for the first long-duration crew. ["Next Space Station mission planned in April," *KSC Countdown*, February 22, 2000.]

◆ Shuttle Endeavour is expected to land after its successful mission to map the planet, but bad weather at KSC could force the ship to land in California. Forecasters say thick clouds and high winds could be a persistent problem this week at KSC, where the ship has two chances to land today, at 4:50 and 6:22 p.m. ["Weather may force shuttle to land in Calif.," *Florida Today*, February 22, 2000, p 1A. & 2A.]

◆ Shuttle Endeavour defied weather forecasters with a twilight touchdown at Kennedy Space Center's Runway 33 of the Shuttle Landing Facility. Main gear touchdown was at 6:22:23 p.m., landing on orbit 181 of the STS-99 mission. Winds canceled an earlier landing attempt. This was the 97th flight in the Space Shuttle program and the 14th for Endeavour, also marking the 50th landing at KSC, the 21st consecutive landing at KSC, and the 28th in the last 29 Shuttle flights. ["Shuttle breezes into Cape at sunset," *The Orlando Sentinel*, February 23, 2000, p A-1 & A-6. "Mission STS-99 scores an "A," lands at KSC on second pass," *KSC Countdown*, February 24, 2000.]

FEBRUARY 23: Chester M. (Chet) Lee, chairman of Spacehab's Astrotech Space Operations unit and a veteran of NASA manned space programs, died on Feb. 23 in Washington, following heart bypass surgery. He was mission director for the Apollo 12-17 manned Moon landings and the Apollo-Soyuz U.S.-Soviet joint manned flight. ["Obituary," *Aviation Week & Space Technology*, February 28, 2000, p 24.]

FEBRUARY 29: Space Gateway Support – the base operations contractor at Kennedy Space Center, Cape Canaveral Air Force Station and Patrick Air Force Base – sent layoff notices to 65 employees Tuesday. The layoffs will go into effect from March 15 to March 17, said Dan Nettuno, Space Gateway Support's human resources director. ["Union may picket over Cape layoffs," *Florida Today*, March 1, 2000, p 10C.]

DURING FEBRUARY: The Spaceport Florida Authority, a state government organization, will finance $300 million in launch infrastructure at Cape Canaveral to support Lockheed Martin Atlas V Evolved Expendable Launch Vehicle operations. Under the unprecedented plan, the authority will own massive Launch Complex 41 and adjacent facilities and lease them back to Lockheed Martin for Atlas V missions. ["News Roundup," *Aviation Week & Space Technology*, February 21, 2000, p 39.]

◆ Command and Control Technologies Corporation (CCT), Titusville, will provide consulting services, product maintenance, and technical support to NASA's John F. Kennedy Space Center for advanced life sciences technology and software. The program at KSC is centered around an isolation chamber in which combinations of crops are grown in a variety of environmental conditions to determine the most productive mix. The chamber, which was originally a pressure chamber for the Mercury and Gemini space programs, was installed at KSC in 1986. ["CCT signs deal to support life sciences project at KSC," *The Brevard Technical Journal*, February 2000, p 1.]

MARCH 2: Two new launch complexes will be christened this week at Cape Canaveral Air Force Station for the next generation of space vehicles to fly from Florida. Boeing and Lockheed Martin will show off the sites for their future Delta 4 and Atlas 5 rockets, respectively, during private ceremonies today and Friday. ["Boeing, Lockheed Martin to showcase new launch sites," *Florida Today*, March 2, 2000, p 1B.]

◆ High-level Disney designers met with the Director of the Kennedy Space Center to discuss a new attraction, the proposed Space Pavilion. The officials agreed that the project could be "mutually complementary" to both Disney and NASA, said Jim Ball, acting head of public affairs for the Kennedy Space Center. ["Epcot gears up for space flight," *The Orlando Sentinel*, March 2, 2000, p B-1 & B-6.]

◆ Endeavour's post-flight inspections were completed last week and the payload bay doors were opened on Friday. Removal of the Shuttle Radar Topography Mission (SRTM) occurred Feb. 29 and it was transferred to the Space Station Processing Facility on March 1. Endeavour's engines are being removed this week. The next planned flight for Endeavour is mission STS-97, the sixth flight to the International Space Station. ["Endeavour starts preparation cycle for its next flight," *KSC Countdown*, March 2, 2000.]

◆ On March 2, 1972, a 550-pound spacecraft called Pioneer 10 lifted off from Cape Canaveral, Fla., atop an Atlas Centaur rocket. The craft succeeded far beyond the dreams fo the scientists who sent it into space. Now, past the 28th anniversary of its launch, it is approaching the 7 billion-mile mark on its journey. That puts it twice as far from the sun as Pluto and, remarkable, it still works. Its cosmic ray detector continued to take measurements and transmit then, with the aid of a tiny, plutonium-powered, 8-watt transmitter, back to Earth. ["Decades later, Pioneer 10 is still going, still calling in," *NASA Current News*, March 14, 2000, p 10-11.]

MARCH 3: Boeing Co. workers unsuccessfully combed a Huntsville landfill looking for two mistakenly discarded parts of the International Space Station valued at $750,000. The nitrogen and oxygen tanks used by astronauts were never found. ["Search fails for space station parts," *Florida Today*, March 4, 2000, p 3A.]

◆ An auction of, among other things, NASA artifacts dating to the 1960s, was held on Merritt Island. The collection was assembled by Charles Bell, retired NASA electrical design engineer who died Feb. 16. Included are two reassembled Atlas rockets and engines and parts from the Saturn, Apollo and Mercury missions. ["NASA collector's passion goes public," *Florida Today*, March 4, 2000, p 1B & 2B.]

◆ Four Boeing Co. engineers at Cape Canaveral walked off the job Friday and joined striking co-workers, company and union officials said. The workers claimed Boeing asked them to do work they were not qualified to do. The engineers provided technical support to the scheduled April 9 launch of a $1 billion missile detection satellite aboard a Titan rocket. ["Boeing engineers walk off," *The Orlando Sentinel*, March 4, 2000, p B-1.]

MARCH 5: Workers damaged an antenna on shuttle Atlantis on Sunday, but NASA officials can't say whether repairs will delay the ship's April flight of mission STS-101 to the International Space Station. The mishap occurred when United Space Alliance workers swung the antenna dish into a work platform while trying to store the equipment for flight. Atlantis is scheduled for launch April 13. ["Workers damage antenna in shuttle Atlantis' cargo bay," *Florida Today*, March 7, 2000, p 1A.]

MARCH 9: Atlantis, Discovery and Endeavour are in the OPF undergoing pre-launch or post-launch preparations. KSC workers are preparing to transfer orbiter Atlantis to the Vehicle Assembly Building as early as March 16. Atlantis is scheduled to fly on both STS-101 and STS-106 missions. Discovery will fly on STS-92; Endeavour will fly on STS-97. ["OPF full as orbiters are prepared for coming launches," *KSC Countdown*, March 9, 2000.]

◆ Students from all over the country are at the KSC Visitor Complex for the FIRST (For Inspiration and Recognition of Science and Technology) Southeast Regional competition March 9-11. Of the 30 high school teams competing in the robot contests, 16 are Florida teams co-sponsored by NASA and KSC contractors. ["NASA-KSC host high school teams in robot contests," *KSC Countdown*, March 9, 2000.]

◆ NASA cut its shuttle work force to dangerously low levels before a problematic shuttle Columbia flight last year, leading to less safety oversight, according to a report by the Space Shuttle Independent Assessment Team. The agency already has started hiring more NASA workers at KSC and plans to increase their ranks by at least 200 during the two years. ["NASA cuts threatened safety in '99," *Florida Today*, March 10, 2000, p 1A.]

MARCH 10: Boeing has offered to reimburse the federal government for two International Space Station tanks, valued at $750,000, that apparently were discarded by mistake at the Marshall Space Flight Center. ["Boeing offers to pay for lost station tanks," *Florida Today*, March 11, 2000, p 2A.]

MARCH 11: This was the final day of the Southeastern Regional For Inspiration and Recognition of Science & Technology (FIRST) Competition in the Rocket Garden at the Kennedy Space Center Visitor Complex. Twenty-nine teams of high school students, aided by engineers from aerospace and technology companies, were competing in a robotics contest for fun and a chance to go to the nationals April 6-8 at Epcot. ["Robot battles offer fun learning," *Florida Today*, March 12, 2000, p 1B & 2B.]

MARCH 13: Two reports were released today. One called the Mars Climate Orbiter Mishap Investigation Board was intended to be a lessons-learned report after the Sept. 23 loss of the Climate Orbiter. The second report, produced by a panel led by Henry McDonald, director of NASA's Ames Research Center in Mountain View, Calif., found NASA had cut its shuttle work force to dangerously low levels. Both reports found that "faster, better and cheaper" is not working the way it should. The space agency will decide what specific actions to take. ["'Faster, better, cheaper' not working, reports find," *Florida Today*, March 14, 2000, p 1A & 2A.]

MARCH 14: A Florida House committee unanimously approved a tax break for spaceflight businesses. The measure (HB-775) would give a sales tax break on property leases for commercial spaceflight companies. It could also help tenants of a proposed $30 million space-research park that lawmakers are pushing for Kennedy Space Center. ["Space-flight businesses get tax break from House panel," *Florida Today*, March 15, 2000, p 1A.]

MARCH 15: Engineers will swap out one of shuttle Atlantis' three main engines because of worries that a pair of defective seals may accidentally have gotten inside. The change-out is not expected to delay the shuttle's mid-April launch on a maintenance mission (STS-101) to the International Space Station. Atlantis is scheduled to be moved from its hangar to Kennedy Space Center's Vehicle Assembly Building early tomorrow, where the engine swap will likely be performed. ["NASA suspects seals on Atlantis are faulty," *The Orlando Sentinel*, March 16, 2000, p A-4.]

◆ U.S. Sen. Bob Graham, D-Miami Lakes, is shaping new legislation that would boost the space industry by enabling commercial companies to lease NASA-owned property for full-time launch and satellite operations. Called the Commercial Space Partnership Act, the legislation is the third piece of a package aimed at helping the space industry in Florida, as well as the rest of the country. If it becomes law, the legislation would have a major impact on Florida, marking the first time non-shuttle commercial launch and satellite operations would be brought to the Kennedy Space Center on a full-time basis. ["Legislation would allow leasing NASA-owned property," *Florida Today*, March 16, 2000, p 4A.]

◆ Last month, NASA received congressional approval to spend $14 million on Russian-built space hardware. One of the Russian-built pieces, the lid-like "pressure dome" worth $4 million, is needed to meet NASA Administrator Dan Goldin's commitment to launch the U.S. built Interim Control Module in December. If the U.S. module is launched, the pressure dome would be needed to allow astronauts to enter the space station. The pressure dome NASA is buying will be built by RSC Energia, formerly a state-owned aeronautics company, and is scheduled to be delivered to Kennedy Space Center in November, said Mike Hawes, deputy associate administrator for space development for NASA. "The last estimate we got from the Russians was that this could be manufactured and delivered to the Cape in November in a time frame to support a December launch of the Interim Control Module," Hawes said. ["Goldin expects crew on station in 2000," *Florida Today*, March 16, 2000, p 5A.]

MARCH 16: U.S. Senator Bob Graham, D-Miami Lakes, spent the day as an iron-worker at Boeing Company's developing Delta 4 launch complex 37B at Cape Canaveral Air Force Station. ["Graham labors at launch pad," *Florida Today*, March 17, 2000, p 1B.]

◆ Construction of the International Space Station could be delayed further because Russian-built components fail to meet NASA safety standards in four key areas, the General Accounting Office reported. ["More delays may be on way for space station," *Florida Today*, March 17, 2000, p 1A.]

◆ NASA is targeting April 17 as the new launch date for STS-101 shuttle Atlantis' 10-day flight to the International Space Station. The agency had been working toward an April 13 liftoff but needs more time for extra work and crew training, officials said. Kennedy Space Center workers are to begin replacing a potentially faulty engine on Atlantis this weekend, clearing the way for moving the shuttle to its launch pad late next week. ["Next shuttle launch," *Florida Today*, March 17, 2000, p 1A.]

MARCH 20: Surviving members of Project Mercury unveiled a statue of Alan B. Shepard Jr., the first American to fly in space, at the U.S. Astronaut Hall of Fame in Titusville. Attending were Mercury Astronauts Scott Carpenter, Gordon Cooper, Wally Schirra and former Sen. John Glenn. ["Space program needs new spirit, astronauts say," *Florida Today*, March 21, 2000, p 1A.]

MARCH 21: The SPACEHAB Double Module, the primary payload on mission STS-101, was transported to the Payload Changeout room at Launch Pad 39A. Space Shuttle Atlantis is scheduled to roll out to Launch Pad 39A Saturday morning starting at 7 a.m. STS-101 will be the 21st flight for Atlantis and the 98th Shuttle flight overall. ["SPACEHAB module at pad, ready for transfer to Atlantis," *KSC Countdown*, March 23, 2000.]

MARCH 23: Managers at NASA's Jet Propulsion Laboratory made public a mishap that seriously damaged a science satellite. The High Energy Solar Spectroscopic Imager, designed to study solar flares, was scheduled for launch in July aboard a Pegasus rocket from Cape Canaveral. That launch is now off until at least January. ["Flubbed NASA test damages $75 million science satellite," *The Orlando Sentinel*, March 24, 2000, p A-13.]

MARCH 25: A Boeing Delta II rocket carrying NASA's IMAGE (Imager for Magnetopause-to-Aurora Global Exploration) spacecraft lifted off at 12:34 p.m. from Vandenberg Air Force Base for a 56-minute ride into orbit. Separation from the launch vehicle took place on schedule and without any reported problems. [Web posted. (2000). NASA launches satellite to study magnetosphere [Online]. Available WWW: http://www.cnn.com/ [2000, March 25].]

◆ Space Shuttle Atlantis rolled out to the launch pad early this morning. The shuttle is tentatively scheduled for a mid-April launch, but the STS-101 mission may be delayed to enable more crew training time. ["State's space industry needs to diversify, lawmaker warns," *Florida Today*, March 26, 2000, p 1A & 2A.]

◆ Florida Space Business Roundtable sponsored a panel discussion of the Florida Space Research Institute (FSRI) to discuss a proposed $30 million space-research park at the Kennedy Space Center. The FSRI, an industry-led organization was set up last year to boost Florida's ability to attract new aerospace companies. ["States space industry needs to diversify, lawmaker warns," *Florida Today*, March 26, 2000, p 1A & 2A.]

MARCH 27: NASA remains undecided on a firm date for the planned mid-April flight of shuttle Atlantis to the International Space Station, officials said. The STS-101 flight, now off at least one day, is tentatively scheduled no earlier than April 18, but could be delayed a few days after the completion of a review now being conducted by shuttle managers, NASA spokeswoman Eileen Hawley said from Houston. NASA originally was aiming for an April 13 flight, but engine work forced a delay to April 17. A dress rehearsal for the mission was moved from this week to next week to give mission commander James Halsell's left ankle more time to heal. Halsell injured the ankle during training at Johnson Space Center in Houston. ["Atlantis delayed at least one day," *Florida Today*, March 28, 2000, p 1A.]

MARCH 29: Roy Bridges briefly turned lobbyist on Wednesday, meeting with House Speaker John Thrasher and House Appropriations Chairman Ken Pruitt to discuss funding for a proposed $50 million space research laboratory at Kennedy Space Center. The laboratory is envisioned as an anchor tenant for a 400-acre space industrial park on adjacent KSC property. Within seven years, the complex is expected to generate 4,000 jobs and another 10,000 jobs in support industries – and millions of dollars in state revenue. ["KSC director lobbies for research lab," *Florida Today*, March 30, 2000, p 10C & 9C.]

MARCH 30: The launch date for mission STS-101 has been rescheduled for Monday, April 24 about 4:15 p.m. EDT in order for the crew to complete training. Landing is scheduled for Thursday, May 4, about 11 a.m. ["STS-101 launch date rescheduled for April 24 about 4:15 p.m." *KSC Countdown*, March 30, 2000.]

◆ NASA Administrator Dan Goldin is taking the blame for last year's botched Mars missions, saying he pushed too hard, cut costs and made it impossible for spacecraft managers to succeed. But Goldin said he will not abandon the National Aeronautics and Space Administration's "faster, better, cheaper" approach. Mission managers will get enough money and personnel to do the job, but there won't be a return to days of big, expensive spacecraft. [Web posted. (2000). NASA Chief Takes Blame [Online]. Available WWW: http://abcnews.go.com/ [2000, March 30].]

◆ The unmanned X-38 – a wedge-shaped craft with no wings and no engine – was dropped from under the wing of a B-52 bomber at 39,000 feet and parachuted safely to the desert floor at NASA's Dryden Flight Research Center. For the first years of the space station, astronauts will have to relay on a three-man Soyuz capsule for emergency escapes. NASA hopes to put a seven-person lifeboat on the space station in 2005 or 2006. ["Tests OK on 'lifeboat' prototype for space," *Florida Today*, March 31, 2000, p 7A.]

◆ A problem with a European satellite is forcing a delay of the inaugural flight of the Lockheed Martin Atlas 3 rocket from Cape Canaveral Air Force Station. The flight had been scheduled for April 14 but will not occur until at least mid-May, Lockheed Martin officials said. ["Satellite problem delays first launch of Atlas 3 rocket," *Florida Today*, March 31, 2000, p 1A.]

DURING MARCH: The Spaceport Florida Authority will facilitate $300 million in private financing to Lockheed Martin Space Systems to develop launch infrastructure for the Atlas V Evolved Expendable Launch Vehicle (EELV) program at the Cape Canaveral Spaceport. The infrastructure, including Launch Complex 41 (LC-41) and adjacent processing facilities, will be owned by the Spaceport Authority and leased to Lockheed Martin to support commercial and government missions. ["Spaceport Authority to finance Florida launch facilities," *The Brevard Technical Journal*, March 2000, p 4.]

◆ NASA Administrator Daniel S. Goldin announced recently that the agency would merge the chief technologist's office with the office of Aero-Space Technology to better focus the agency's strategy for maintaining its long-term technology base. Samuel Venneri, chief technologist, will retain that position while becoming Aero-Space Technology's associate administrator. ["Venneri to head merged technology, aero-space office," *The Brevard Technical Journal*, March 2000, p 5.]

◆ The Cape Canaveral Technical Society, during its Engineers' Week Banquet, presented the CCTS Engineering Achievement Award to Dr. Pedro J. Medelius, who was recognized for the revolutionary invention and testing of a low-cost method of locating the point of attachment of lightning strikes to pad structures. This dynamic new technology is called the Universal Signal Conditional Amplified (USCA). Dr. Medelius led a crew of 15 scientists, engineers and technicians at the Data Acquisition Laboratory at Kennedy Space Center. ["Engineering Award," *KSC Countdown*, March 28, 2000.]

◆ The liftoff of Atlantis in April on logistics flight to the International Space Station will debut the shuttle's new glass cockpit, one of the most significant program upgrades since the first launch 19 years ago. The new system will give shuttle commanders and pilots a major increase in situational awareness. This would be especially important for any launch abort cases, where additional unambiguous information, displayed more clearly, could mean the difference between life or death to a crew. Astronauts USAF Col. James Halsell and USAF Lt. Col. Scott Horowitz are to pilot Atlantis on STS-101, its first mission of the new multifunction electronic display system (MEDS) cockpit. ["MEDS Glass Cockpit To Enhance Shuttle Safety," *Aviation Week & Space Technology*, March 6, 2000, p 54-56.]

◆ NASA has taken a lot of heat from Republicans for a $75 mission effort to place a satellite at the L1 point so it could stare continuously at the home planet and provide real-time imagery on the Internet. Called Triana, it is a pet project of Vice President Al Gore and scheduled to fly early next year. Congressional appropriators put the project on hold last year and ordered NASA to ask the National Research Council what it thought. Last week, the NRC reported that Triana would complement other Earth science missions and is worth the cost. ["State Way," *Aviation Week & Space Technology*, March 13, 2000, p 23.]

◆ A NASA-funded effort by Stanford University to test Einstein's theory of relativity faces $70 million in cost overruns and an additional 6-month launch delay. Originally slated for a 1999 launch, the Gravity Probe B is designed to measure how Earth's mass warps space-time. April 2002 is more likely, stated Rex Geveden, program manager at NASA's Marshall Space Flight Center in Huntsville, Alabama. ["Web posted. Andrew Lawler. (2000). Einstein Probe remains Earthbound [Online]. Available WWW: http://www.sciencemag.org/ [2000, March 10].]

APRIL

APRIL 2: The next group of astronauts to visit the International Space Station will be in town this week, practicing for an April 24 launch on shuttle Atlantis. The STS-101 crew is to run through a countdown dress rehearsal at Kennedy Space Center on Thursday and Friday, when they will put on their launch suits and get strapped into their seats aboard the spaceship. Originally scheduled for last week, the dress rehearsal wad delayed so Atlantis Commander Jim Halsell could recover from a sprained ankle and take part in the activities. ["Shuttle crew to rehearse this week for launch," *Florida Today*, April 3, 2000, p 1B.]

◆ Federal workers who help explore space are among the most satisfied with their jobs according to an Office of Personnel Management survey. The highest satisfaction scores went to the National Aeronautics and Space Administration with 78 percent. ["Workers at NASA score highest in federal job-satisfaction survey," *NASA Current News*, April 3, 2000, p 37.]

APRIL 4: Boeing Space and Communications Group has signed a $26.3 million contract modification with NASA for work on the International Space Station. The modification covers planned changes to the baseline assembly sequence, including moved launch dates, deleted and added U.S. Space Shuttle flights to the Station, and revisions to the Multi-Increment Manifest (MIM) for the orbiting facility. ["Boeing Space and Communications," *NASA Current News*, April 14, 2000, p 30.]

APRIL 5: The STS-101 crew is at KSC for TCDT today and tomorrow. The crew of seven comprises Commander Jim Halsell, Pilot Scott Horowitz, and Mission Specialists Mary Ellen Weber, Jim Voss, Jeff Williams, Susan Helms and Yuri Usachev. ["TCDT under way with STS-101 crew," *KSC Countdown*, April 6, 2000.]

◆ Spacehab, Inc. has won a $21.6 million modification to its mission support contract with NASA to provide a pressurized double module and a Russian-built cargo pallet for the U.S. Space Shuttle mission that will outfit the Russia's Zvezda Service Module after it is launched this summer. The modification to Spacehab's Research and Logistics Mission Support (REALMS) contract covers the STS-106 mission, scheduled to fly in August after Zvezda is delivered to the growing Station in July. ["Spacehab gets $21.6 million for August Space Station supply mission," *NASA Current News*, April 14, 2000, p 22.]

APRIL 6: Shortly after NASA managers firmed up an April 24 flight date for the shuttle Atlantis earlier this week, mechanical problems were discovered that could shove the mission into May. Problems with a power unit affecting the shuttle's landing mechanism were discovered by an engineer April 5th. If the work can be done at the launch pad, the current date might hold, said NASA spokesman Bruce Buckingham. But if shuttle managers are forced to remove the Atlantis from the launch pad to do the work elsewhere at Kennedy Space Center, the delay could be several weeks, Buckingham said. ["Mechanical problems could delay launch," *Florida Today*, April 7, 2000, p 1A.]

◆ A planned May 31 launch of a communications satellite aboard a Boeing Co. Delta 3 rocket has been delayed until sometime in October, Boeing officials announced Thursday.

["Web posted. (2000). Delta 3 launch delayed until October [Online]. Available WWW: http://www.flatoday.com/ [2000, April 07].]

◆ Pamela Bookman, a NASA/Kennedy Space Center employee, recently joined the United States Space Foundation's Space Technology Hall of Fame. Bookman was officially inducted on April 6 at the 16th National Space Symposium in Colorado Springs, Colo. Joining her were Richard H. Beck and Daniel A. Drake, employees of NASA contractor United Space Alliance. The three co-developed an advanced lubricant used for the crawler-transporters that deliver the Space Shuttle to the launch pad. ["Pamela Bookman," *KSC Countdown*, April 13, 2000.]

APRIL 7: NASA managers are gearing up to replace a power unit that affects the landing mechanism on shuttle Atlantis, officials said Friday. A replacement part from shuttle Columbia is being flow to Kennedy Space Center from Palmdale, Calif, where Columbia is undergoing an extensive overhaul before the STS-101 mission. If the work is done at the launch pad, the flight likely will proceed as scheduled April 24, but if the work must be done elsewhere at KSC, the flight might be postponed until May. ["Power unit to be replaced on shuttle," *Florida Today*, April 8, 2000, p 1A.]

APRIL 8: The Kennedy Space Center Visitor Complex will eliminate reduced-price options and free access to parts of the park, and instead will charge visitors a flat admission price like those at major theme parks. The new ticket prices -- $24 for adults and $15 for children ages 3 to 11 -- are intended to "take us to the next level" of the tourism industry, said Rick Abramson, president and chief operating officer of Delaware North Parks Services of Spaceport Inc. The private company operates and manages the park and Space Center bus tours for NASA. It drew 2.83 million people there last year. Delaware North officials said the new ticket price structure is intended to encourage visitors to see all the attractions and spend more time at the complex. ["KSC complex to end free access," *Florida Today*, April 7, 2000, p 1A & 3A.]

◆ Former astronaut and NASA manager Loren Shriver has been named deputy program manager of operations for United Space Alliance, the main contractor overseeing NASA's space shuttles. Shriver served as Kennedy Space Center's deputy director in charge of launch and payload processing. Shriver starts in his new position Monday, April 10th. ["Ex-astronaut named manager of Space Alliance," *Florida Today*, April 9, 2000, p 1E.]

APRIL 11: NASA will start work today to replace faulty equipment in shuttle Atlantis' landing system. The shuttle remains scheduled for an April 24 launch to the International Space Station, but an on-time liftoff for mission STS-101 depends on NASA replacing the broken component – a 340-pound power drive unit – with the shuttle in a vertical position. If it can't be done for some reason, the ship would have to be rolled back to its hangar at KSC, and the flight would be delayed until May. "They've never replaced this power unit at the pad before, but they've done a lot of work assessing it, and they're very comfortable with it," KSC spokesman Joel Wells said. ["Shuttle part to be replaced at pad," *Florida Today*, April 11, 2000, p 1A.]

◆ NASA is studying the process used to provide press credentials for shuttle launches. Bruce Buckingham, NASA's news chief at Kennedy Space Center, two other NASA public affairs officers and three journalists make up a panel to oversee all aspects of the information business. The new press-credential policy should be in place sometime after the April 24 launch of shuttle Atlantis (STS-101), Buckingham said. ["Agency plans new policy on press credentials," *Florida Today*, April 12, 2000, p 3A.]

APRIL 14: The Cryogenics Testbed Facility, a new venture in technology and research collaboration, was unveiled on April 14 at a ribbon-cutting ceremony. The facility, jointly managed by NASA and Dynacs Engineering Company, was created to better apply cryogenics to lives in the field of medicine, biology, food, computers, industry, rocket propulsion and the spaceports of the future. KSC's Cryogenic Testbed Facility comprises the *Cryogenic Test Laboratory, Liquid Nitrogen Flow Test Area, Hazardous Test Area* and the *Launch Equipment Test Facility*. ["New Facility to Open," *KSC Countdown*, April 6, 2000.]

APRIL 18: Over the weekend, engineers conducted a Frequency Response Test on the replacement Power Drive Unit (PDU) in Atlantis and confirmed the Shuttle's entire hydraulic system is functioning normally. Also, the Auxiliary Power Unit (APU) flex hose was replaced successfully. The STS-101 launch remains on track for April 24 at about 4:15 p.m. ["Shuttle tests on PDU okay, launch on target for April 24," *KSC Countdown*, April 18, 2000.]

◆ The planned launch of a Delta 2 rocket carrying a military navigation satellite from Cape Canaveral Air Force Station has been rescheduled for April 22nd. In a statement, officials said a "technical issue with ground support equipment that provides electrical power" to the satellite caused the postponement. The launch window is from 11:01 to 11:30 p.m. ["Rocket launch delayed," *Florida Today*, April 22, 2000, p 1A.]

APRIL 20: Kennedy Space Center Visitor Complex has started selling 12-month passes to the park as part of a new ticket-pricing plan that started April 8. The passes allow visitors unlimited visits to all park attractions and the KSC bus tour. Under the offer announced, Brevard resident can buy the passes at a reduced cost. In addition to annual passes, Delaware North offers NASA and related civil service employees and retirees free admission to the limited-access areas of the park, and offers discounts on the full admission price of tickets bought at the Space Center's NASA Exchange stores. Delaware North's new pricing policy eliminated the reduced-price options and free admission to parts of the park. ["KSC offers discounted annual passes to Brevard residents," *Florida Today*, April 21, 2000, p 1A.]

APRIL 21: The space shuttle Atlantis will fly with an updated "glass cockpit" when it lifts off Monday (April 24) on the STS-101 mission to service the fledgling International Space Station, NASA officials said. The new cockpit replaces dozens of conventional gauges and cathode-ray tube displays with 11 flat-panel color screens that provide easier pilot recognition of key functions, NASA said. The cockpit is 75 pounds lighter and uses less power than its predecessor and is expected to become standard on all shuttles by 2002. Web posted. (2000). Shuttle set for liftoff with 'glass cockpit' [Online]. Available WWW: http://www.cnn.com/ [2000, April 21].]

◆ NASA continues the countdown toward a Monday launch of shuttle STS-101/Atlantis, as the astronaut crew and Kennedy Space Center workers make final preparations for the flight. The three-day countdown started soon after Atlantis' seven-member crew flew into KSC from Houston. NASA officials say everything is on track for liftoff between 4:15 and 4:20 p.m. Forecasters say there is a 70 percent chance of favorable weather at launch time, with the possibility of clouds causing a problem. The flight originally was scheduled for April 13 but was delayed twice -- once for engine problems and once to give Halsell's injured ankle time to heal. ["Shuttle countdown under way," *Florida Today*, April 22, 2000, p 1A.]

APRIL 22: The launch of a Delta 2 rocket was called off because of potential problems with the $42 million navigation satellite the rocket was to carry. This is the second delay in two days. A new launch date has not been set for the rocket, which was to have left Cape Canaveral Air Force Station between 11:01 and 11:30 p.m. The Delta rocket, built by the Boeing Co., was to launch the Navstar Global Positioning system 2R-4 for the Air Force. ["Air Force scraps Delta 2 launch for 2nd time," *Florida Today*, April 23, 2000, p 2A.]

◆   Gus Grissom's Mercury capsule, pulled from the Atlantic Ocean last summer, will leave on a three-year tour next week. It is scheduled for display June 17 through September 17 at Kennedy Space Center's Visitor Center. ["Gus Grissom's capsule leaves for 3-year tour," *Florida Today*, April 23, 2000, p 8E.]

◆   During the STS-101 mission, Atlantis' astronauts will use a revamped ventilation system, portable fans and earplugs to protect their health while working on the International Space Station. Some of the measures are meant to prevent the headaches, dizziness and nausea that struck members of the last crew to visit the outpost in June 1999. The brief bouts of illness arouse while the astronauts worked inside the station for a long time but dissipated soon after they floated back onto shuttle Discovery during the STS-96 mission. ["Vents, fans to prevent sickness in astronauts," *Florida Today*, April 23, 2000, p 3A.]

APRIL 23: Teresa Annulis, United Space Alliance, is only the second woman to take the seat as orbiter test conductor inside Kennedy Space Center's Launch Control Center. Annulis will be the last person on the launch team to talk with the STS-101 crew before liftoff. The orbiter test conductor oversees all the work done on a shuttle from the moment it lands to liftoff. ["Titusville woman guides crew," *Florida Today*, April 24, 2000, p 8A.]

APRIL 24: High winds forced the scrub of the STS-101 launch minutes before liftoff. Launch was called off at the nine-minute mark of the countdown after wind gusts exceeded the safety limit. ["Launch of STS-101 scrubbed due to high winds – launch planned for today," *KSC Countdown*, April 25, 2000. Web posted. (2000). Winds Delay Launch Again. [Online]. Available WWW: http://www.abcnews.com/ [2000, April 25].]

APRIL 25: For the second day in a row, high wind forced NASA to delay the launch of STS-101 space shuttle Atlantis on a mission to repair the International Space Station. Gusts as high as 37 mph, close to gale force, swept the pad as NASA counted down toward an afternoon liftoff. With little improvement expected, launch managers called a halt and said they would try again Wednesday. [Web posted. (2000). Winds Delay Launch Again. [Online]. Available WWW: http://www.abcnews.com/ [2000, April 25].]

◆   United Space Alliance, prime contractor to NASA for Space Shuttle operations, recently introduced DeepWorker 2000, a one-man submarine designed to help recover the expended solid rocket boosters after they splash down in the Atlantic Ocean. USA will demonstrate the capability of its sub in post-flight operations during the STS-101 mission. ["One-man submarine to demo ability on SRB retrieval from ocean after launch," *KSC Countdown*, April 25, 2000.]

◆   Florida House and Senate budget negotiators have tentatively agreed to spend $10 million, and possibly more, on an industrial park at Kennedy Space Center that supporters hope will help preserve Florida's dominance in the industry. The House and Senate aren't expected to give final approval on a budget until early next week, but both chambers have agreed to pump $10 million into the project this year and again next year. The money would be used to build a high-tech laboratory at Kennedy Space Center that would mostly be used to process life-science experiments for the International Space Station. The state would build the facility and lease it to NASA. ["Legislature approves industrial park at KSC," *Florida Today*, April 26, 2000, p 1B.]

APRIL 26: STS-101 was scrubbed again, this time due to weather conditions at the overseas TransAtlantic Landing (TAL) sites. At least one site must be clear in case problems during liftoff force an emergency landing. Rain soaked two sites in Spain. A third landing strip in Morocco, had winds gusting at 16 knots by launch time. The flight now has to compete for dates on the Air Force Eastern Range, which runs the vast tracking network that monitors all shuttle and rocket flights from Brevard County. ["No April launch for Atlantis," *Florida Today*, April 27, 2000, p 1A.]

APRIL 27: The NASA/NOAA weather satellite GOES-L is scheduled to launch May 3, from Cape Canaveral Air Force Station's Pad 36A on an Atlas IIA rocket. ["ELV Update," *KSC Countdown*, April 27, 2000.]

APRIL 28: Shuttle Atlantis, STS-101, will have to wait until May 18 for its next chance to fly so a hurricane satellite can keep its planned launch attempt next week, NASA officials announced today. NASA had considered delaying that flight so Atlantis could make a fourth launch attempt next week but in the end NASA officials decided to launch the NOAA satellite. Atlantis is set for liftoff from Kennedy Space Center at 6:33 a.m. May 18. ["Atlantis' next launch try May 18," *Florida Today*, April 29, 2000, p 1A.]

DURING APRIL: NASA's Near Earth Asteroid Rendezvous (NEAR) spacecraft conducting the first-ever close-up study of an asteroid will be renamed to honor Dr. Eugene M. Shoemaker, a legendary geologist who influenced decades of research on the role of asteroids and comets in shaping the planets. ["NASA renames NEAR spacecraft," *The Brevard Technical Journal*, April 2000, p 4.]

◆   In a recent "topping-off" ceremony, Lockheed Martin Space Systems Company positioned the top steel beam in place on its new 300-foot tall Atlas V Vertical Integration Facility (VIF) at Cape Canaveral Air Force Station Space Launch Complex 41 (SLC-41). . ["Lockheed Martin tops off Atlas V vertical integration facility at SLC-41," *The Brevard Technical Journal*, April 2000, p 5.]

MAY

MAY 1: The 37th Space Congress, Space Means Business in the 21st Century, opened with a keynote address from space.com CEO Lou Dobbs. "We think this will be one of our best conferences every," event chairman Charles Murphy said. "This also is a special year for us because it marks the 50th anniversary of the first rocket launch from Cape Canaveral on July 24, 1950." ["Business key for Space Congress," *Florida Today*, May 1, 2000, p 1E & 5E.]

MAY 3: A Lockheed Martin Atlas 2A rocket successfully carried the Geostationary Operational Environmental Satellite-L (GOES-11) into space from Cape Canaveral Air Force Station at 3:07 a.m. EDT. No problems were reported during the launch that marked the 49th straight success for the Atlas rocket family dating back to 1993. [Justin Ray. (2000). U.S. has a new weather satellite circling Earth. [Online]. Available WWW: http://www.cnn.com/ [2000, May 3].]

MAY 4: Kennedy Space Center Visitor Complex hosted a job fair at the Early Space Exploration Building. ["KSC Visitor Complex hosts job fair Saturday," *Florida Today*, May 5, 2000, p 9C.]

◆ Kennedy Space Center's National Day of Prayer observance was held on May 4, in the O&C Building, Mission Briefing Room. [Memoranda (NASA). Roy D. Bridges, Jr. Subject: "National Day of Prayer 2000," May 1, 2000.]

MAY 5: The Florida legislature passed a budget that included $10 million for a space research laboratory at Kennedy Space Center. The space research laboratory would be built by the state and leased by NASA. Also included was $4 million to build an access road that would serve the KSC Visitor Complex and the laboratory. ["Space projects fare well during 2000 session," *Florida Today*, May 6, 2000, p 3A.]

MAY 8: Kennedy Space Center re-organized to focus on safe operations of the Space Shuttle, Space Station, Expendable Launch Vehicle programs and customers, and spaceport technology development. Highlights of the new structure include creating a Spaceport Engineering and Technology organization to focus on spaceport technology and development and project management; a Spaceport Services group to service the needs of the center's internal and external customers; and establishment of an External Relations and Business Development team to create a "one-stop shopping" for new customers and improve internal and external communications. Roy Bridges, KSC's Center Director, said the number one driving force for the change is that a reduced workforce has left NASA with critical skill shortages. Through the reorganization Bridges said, "We hope to attract, develop and retain a highly competent, diverse, agile and flexible workforce." Other factors leading to the reorganization include redundant functions within the current organizational structure, an excessive number of internal customer handoffs, limited flexibility for employees, and an inconsistent alignment with the KSC Roadmap. ["Kennedy Space Center Rolls Out New Organizational Structure," *KSC Press Release #46-00*, May 8, 2000.]

◆ After three failed missions from Cape Canaveral Air Force Station during the past two years, a massive 19-story Titan 4B rocket was successfully launched carrying a missile-detection satellite. The first failure occurred on August 12, 1998, the next failure occurred on April 9, 1999 and the last failure from the Cape was on April 30, 1999. The next flight of a Titan 4B is scheduled for July from Vandenberg Air Force Base. The next flight of the Titan 4B from the Cape is planned for the fall. ["Titan launch a success," *Florida Today*, May 9, 2000, p 1A & 2A.]

MAY 10: A Delta 2 rocket sent a military navigation satellite, Navstar Global Positioning system 2R, into space from Cape Canaveral Air Force Station at 9:48 p.m. ["Cape celebrates third launch success in a week," *Florida Today*, May 11, 2000, p 1A.]

MAY 15: The countdown for STS-101 began at 9:30 a.m. Launch remains scheduled for 6:38 a.m. May 18. ["STS-101 crew arrived, countdown underway," *KSC Countdown*, May 16, 2000.]

◆ The Air Force and NASA are working to resolve a tracking-station radar problem that forced the cancellation of the inaugural launch of Lockheed Martin's Atlas 3A rocket. The malfunctioning radar at a tracking station in Bermuda ended the Atlas launch attempt about 6 p.m., a few minutes after the launch window opened. ["Radar flaw delays launch," *Florida Today*, May 16, 2000, p 1A.]

MAY 16: The second canceled launch attempt of the new Atlas 3A in two days forced NASA to move the planned liftoff of space shuttle Atlantis, STS-101, from Thursday (May 18th) to Friday (May 19th). Launch time from Kennedy Space Center is set for 6:12 a.m. Because of the new launch date, Atlantis' journey to the International Space Station will be cut from 11 days to 10, but the mission will not be adversely affected, officials said. The shuttle is expected to dock with the station early Sunday, undock in the late afternoon May 26 and return to KSC early May 29 – the same landing date as before. Under an agreement with Air Force launch-range officials, NASA agreed to give the Atlas three opportunities to fly this week. ["Atlas 3A delays shuttle," *Florida Today*, May 17, 2000, p 1A.]

MAY 17: The STS-101 mission with space shuttle Atlantis is gearing up for a 6:12 a.m. liftoff Friday (May 19th) from Kennedy Space Center. The Atlas launch from Cape Canaveral Air Force Station was canceled after a series of technical problems with the rocket, none of which appeared to be major, Lockheed Martin spokeswoman Julie Andrews said. The Atlas came within less than 30 seconds of making its inaugural flight before technical problems with the rocket arose. ["Shuttle set for liftoff Friday, *Florida Today*, May 18, 2000, p 1A & 2A.]

MAY 19: Space Shuttle Atlantis, STS-101, lifted off from Launch Complex 39A at 6:11 a.m. It was the 98th time that NASA's shuttle fleet has been sent into space since the program started in 1981. "It was awesome...What a beautiful time of day to launch," said Dave King, NASA's director of shuttle processing at KSC and a 17-year space-agency veteran. Etched against a clear pink and blue Cape sky, the shuttle was visible for several minutes as it headed toward the International Space Station 207 miles above Earth. The launch not only was something special to watch, it was carried out with no problems. ["Shuttle bound for station," *Florida Today*, May 20, 2000, p 1A & 3A.]

◆ Ed Stone, director of NASA's Jet Propulsion Laboratory, will retire next year it was announced. ["NASA director Stone retires," *The Orlando Sentinel*, May 20, 2000, p A-6.]

MAY 20: The inaugural launch of a commercial Atlas 3A rocket was canceled for the fourth time in a week. This time, boaters were to blame. Air Force and Coast Guard officials said that as many as 70 boats were in a danger zone off Port Canaveral during the time the rocket was to be launched from Cape Canaveral Air Force Station. A few minutes before the window closed at 7:57 p.m., the boaters had been cleared and Lockheed Martin was given permission to launch. Moments later, however, a technical problem finished the launch attempt. ["Atlas scrubbed 4th time," *Florida Today*, May 21, 2000, p 1A & 2A.]

MAY 21: NASA commissioned a study to determine whether the space center is adequately prepared for a major hurricane. The shuttles will remain at KSC because even though Hurricane Floyd came dangerously close to the Cape in 1999, the odds of a major storm strike remain historically low. In the past 40 years, the space center has never been hit by a Category 3 (hurricane) or higher. KSC has taken steps to ensure the space ships are better protected from wind and flooding than they were during Hurricane Floyd. One of those steps is to move all the shuttles into the Vehicle Assembly Building, which is designed to handle winds up to 125 mph. However, flooding remains a big concern. As a result, all four bays of the VAB have been cleaned out and the cranes readied to suspend the entire shuttle fleet several stories above the ground. Also the center's skeleton crew of 105 people usually stay in the astronauts' barracks and other buildings. The study has shown that the best place for the emergency workers to be during a storm is on the second or third floor of the Launch Control Center. ["KSC still plans to keep shuttles at Cape in storms," *Florida Today*, May 22, 2000, p 1B & 2B.]

MAY 23: A House subcommittee voted Tuesday to give NASA a $13.7 billion budget in the next fiscal year, $321.7 million less than the Clinton administration's request. But the amount approved by the House Appropriations subcommittee on independent agencies, including NASA, was $113 million more than the budget Congress gave the space agency for the current fiscal year. The panel's vote, which starts NASA's drawn-out budget process, is a far cry from this point last year when the same panel approved a bill to cut the agency's funding by $1 billion. President Clinton's request in February asked for $14 billion, a 3 percent increase from this year's level and the agency's first raise in seven years. Funding for the human space flight portion of NASA's budget, which includes operating the shuttle fleet and continued spending on the International Space Station, was approved at the level the administration requested. Subcommittee members cut the most -- $290 million – from NASA's Space Launch Initiative, a variety of projects including the second-generation reusable launch vehicle. The bill now goes to full committee and then to the House floor. The Senate has yet to take up its version. ["House panel sets NASA budget figure," *Florida Today*, May 24, 2000, p 1A & 2A.]

MAY 24: An Atlas 3A rocket lifted off at 7:10 p.m. from pad 36A carrying a European communications satellite in space. The launch occurred after the Air Force and Coast Guide were able to shoo away about 20 boats from a launch-debris danger zone off Port Canaveral. An aircraft also infiltrated the zone. This launch marked the first time a Russian rocket engine, RD-180, has powered an American booster. ["Atlas rides into space on Russian firepower," *The Orlando Sentinel*, May 25, 2000, p A-15. "Atlas 3A finally lifts off," *Florida Today*, May 25, 2000, p 1A & 5A.]

MAY 25: The crew of Space Shuttle Atlantis is completing mission activities to repair equipment in the Zarya control module of the International Space Station. Along with batteries, the crew has installed 10 new smoke detectors, four cooling fans, and new cabling for the module's central computer. Plus the astronauts have transferred such items as exercise equipment and computer printers to the Station for use by future crews. Atlantis is scheduled to land at the Shuttle Landing Facility on Monday, May 29, at 2:19 a.m. ["Mission nearly complete for Atlantis crew," *KSC Countdown*, May 25, 2000.]

◆ The next launch of a NASA payload is the Tracking and Data Relay Satellite known as TDRS-H. It is scheduled to be launched June 29 aboard an Atlas IIA/Centaur rocket currently being erected at Launch pad 36A, Cape Canaveral Air Force Station. The satellite is scheduled to arrive at the Shuttle Landing Facility on Friday, May 26 to begin processing. ["ELV Update," *KSC Countdown*, May 25, 2000.]

◆ The Kennedy Space Center Visitor Complex annual Space Days 2000 explores both the scientific and science fiction of life possibly existing on other planets. Activities begin May 27, 2000 and continue through Monday, May 29. Special guests include Nichelle Nichols, from the original Star Trek series and cast members from the popular series the X-Files. An actual SETI scientist will conduct presentations and answer guests' questions. Also on hand will be Apollo astronauts T. K. Mattingly and Walt Cunningham, as well as Shuttle astronauts Mike Mullane, Dr. Story Musgrave and Rick Searfoss. ["Space Days at the Visitor Complex," *KSC Countdown*, May 25, 2000.]

◆ A national panel that includes four former Air Force generals has embarked on a six-month study to determine whether a separate space force should be created. Called the "Commission to Assess United States National Security Space Management and Organization," it is also known as the Space Commission. ["Panel to study need for separate space force," *Florida Today*, May 26, 2000, p 1A & 3A.]

◆ The launch of Russian-built living quarters, Zvezda module, for the International Space Station is set for early July, officials said. The launch will take place from the Baikonur cosmodrome in Kazakstan. ["Station module set for July liftoff," *The Orlando Sentinel*, May 26, 2000, p A-4.]

MAY 26: Shuttle Atlantis' planned return to Kennedy Space Center early Monday could be delayed if firefighters don't contain a wildfire that is blowing thick smoke across the KSC runway. The smoke reduced visibility to between three and four miles at the landing site Friday. NASA requires at least five miles of visibility to land the shuttle at night. The fire is three miles southwest of the runway. ["Smoke may delay shuttle landing Monday," *Florida Today*, May 27, 2000, p 1A.]

MAY 27: The seven crew members of shuttle Atlantis, STS-101, are scheduled to arrive at Kennedy Space Center early Monday (May 28th) after a mission to the International Space Station. Touchdown for Atlantis after the 10-day mission is set for 2:20 a.m. Forecasters are calling for showers, clouds and crosswinds. Also smoke from wildfires in the area has been blowing across the KSC runway. Atlantis will have two opportunities to land Monday: 2:20 a.m., then at 3:56 a.m., KSC spokesman Joel Wells said. If a landing is not possible on Monday, two attempts could be made Tuesday: the first at 1:16 a.m. and the second at 2:51 a.m. ["Weather may delay shuttle's return," *Florida Today*, May 28, 2000, p 1A.]

◆ The Compton Gamma Ray Observatory that was deployed by shuttle Atlantis on April 15, 1991, will return to Earth in a controlled re-entry into the Pacific Ocean ending a successful nine-year space mission. ["NASA set to crash Compton observatory," *Florida Today*, May 28, 2000, p 2A.]

MAY 29: Space Shuttle Atlantis landed, Monday, Memorial Day, at 2:20 a.m. EDT, returning from the International Space Station after 9 days, 20 hours and 9 minutes. The landing occurred on orbit 155, with Atlantis touching down on Runway 15 ending the STS-101 mission. The crew traveled 4,076,241 miles during the mission. This marked the 14th nighttime landing in Shuttle history and the 22nd consecutive landing at KSC. ["Mission complete, Atlantis and crew land on orbit 155," *KSC Countdown*, May 30, 2000.]

MAY 30: A post-landing inspection on the shuttle Atlantis revealed the lower surface of the ship "had taken 64 debris hits," with only 18 measuring an inch or larger. "The debris hits were not significant," said Joel Wells, NASA spokesman. Included in the minor damage were scratches and dents on the thermal tiles of Atlantis' right wing. The damage apparently was caused by ice falling from an external fuel tank. The left wing was also struck by debris during the flight, but the damage was not noticed until Atlantis came home. ["NASA eyes next mission as Atlantis crew departs," *Florida Today*, May 31, 2000, p 1B.]

MAY 31: A critical Russian-built segment of the International Space Station has passed all tests in plenty o time for its July launch, the Russian Space Agency said. Launch of Zvezda Service Module, which is intended to house the station's crew, has been delayed for more than two years. ["Russian service module set for summer launch," *Florida Today*, June 1, 2000, p 1A.]

DURING MAY: The Kennedy Space Center Visitor Complex has revamped the "Blue Tour" bus ride to "Cape Canaveral: Then and Now." Stops on the tour include: Pad 34, Pad 19, Mercury 7 Astronauts Memorial, Pad 14, Pad 13, Cape Canaveral Lighthouse, Minuteman bunkers, Pad 18, Mercury crew quarters, Air Force Space Museum and Complex 5/6 Blockhouse. ["Tour space history 'Then and Now'," *Florida Today*, May 28, 2000, p 1D & 2D.]

◆ According to Florida Today, motion picture and television projects on Florida's Space Coast have directly injected $37 million into the local economy. Most of these funds were generated by big-budget space-related productions. ["Space-related film & TV projects boost Space Coast economy," *The Brevard Technical Journal*, May 2000, p 5.]

◆ Lockheed Martin, Pratt & Whitney and the Air Force will press the Transportation Dept. and Congress for tougher measures to punish the kind of boating incursions that result in costly launch scrubs like that of the new Atlas III on its fourth launch attempt May 20. The Atlas finally flew May 24, but not without being delayed again by a boater and a light aircraft. Under current law, USAF and the Coast Guard can do little beyond issuing

Notices to Mariners and other warnings. For commercial launches, boaters entering the zone can't even be fined. ["Boats, Planes and Rockets," *Aviation Week & Space Technology*, May 29, 2000, p 21.]

◆ NASA is approaching the selection of two $56-million Phase I International Space Station Crew Rescue Vehicle (CRV) study contracts. Current NASA plans call for delivery of the first CRV to the Kennedy Space Center in January 2005. ["CRV competition," *Aviation Week & Space Technology*, May 22, 2000, p 19.]

JUNE

JUNE 1: Atlantis is in Orbiter Processing Facility bay 3 undergoing post-flight inspections and routine vehicle safing. The orbiter is scheduled to fly back to back missions to the International Space Station with STS-106, targeted for launch Sept. 8. ["Atlantis in OPF – will be orbiter on next mission in September," *KSC Countdown*, June 1, 2000.]

◆   NASA has signed an agreement with Dreamtime Holdings Inc., to wire the shuttle fleet and International Space Station for digital television, the Internet and more. Dreamtime will also make tens of millions of NASA photographs, documents, blueprints and video clips available on the Internet. The multimedia company will operate from NASA's Ames Research Center in Mountain View, Calif. The deal will not immediately affect NASA TV employees at Kennedy Space Center. The agency's television programming will remain largely intact for now, NASA officials said. ["NASA deal to expand public's view," *Florida Today*, June 2, 2000, p 1A & 2A.]

JUNE 3: Keith Cowing, a former NASA employee and editor of an independent web site called NASA Watch, received press accreditation from NASA. This is the first time that NASA has accredited the Internet reporter. ["NASA accredits Internet journalist," *Florida Today*, June 4, 2000, p 1B.]

JUNE 6: A federal grand jury has indicted a jailed man on charges he made bomb threats to workers at Kennedy Space Center, the U.S. Attorney's Office announced. NASA criminal investigation officials handled the case. ["Jury indicts man in KSC threats," *Florida Today*, June 6, 2000, p 1B.]

JUNE 7: The KSC Visitor Complex will introduce new discounts for Brevard residents, and their guests, to offset a recent change in admission policies. "We wanted to do what was right for Brevard County residents," said Dan LeBlanc, marketing director for Delaware North Parks Services of Spaceport Inc., the private company that operates and manages the park and Space Center bus tours for NASA. ["Brevard residents get discount at KSC gates," *Florida Today*, June 8, 2000, p 1A & 2A.]

JUNE 8: The Safe Haven project that began in August 1999 is nearing completion. The project strengthens KSC's readiness for hurricane season by expanding the VAB's storage capacity. A new high bay 2 will allow NASA to preassemble stacks and still have room in the VAB to pull a Shuttle back from the pad if severe weather threatens by moving a stack into high bay 2, freeing up bay 1 or bay 3. Last week, the road and high bay 2 were tested with a mobile launcher platform (MLP) atop a crawler-transporter. ["Safe Haven at VAB undergoing tests," *KSC Countdown*, June 8, 2000.]

JUNE 11: The crew on mission STS-92, the fifth flight to the International Space Station, is at KSC this week to take part in Crew Equipment Interface Test (CEIT) activities. The crew of seven comprises Commander Brian Duffy, Pilot Pam Melroy, and Mission Specialists Koichi Wakata, Leroy Chiao, Jeff Wisoff, Michael Lopez-Alegria and Bill McArthur. Wakata is with the Japanese space agency. STS-92 will be the 100th flight in the Shuttle program, aboard Discovery. ["STS-92 crew at KSC to look over payload on 100th mission," *KSC Countdown*, July 11, 2000.]

JUNE 14: After two failures in its first two launch attempts during the past two years, a test launch of a Boeing Co. Delta 3 rocket is scheduled for August 19th. ["Boeing plans test launch to prove efficiency of Delta 3," *Florida Today*, June 15, 2000, p 1A.]

◆ Potential flammable contamination in the form of oily droplets, has been found in emergency oxygen units on all 12 spacesuits used by NASA's shuttle astronauts. Engineers will clean and add new filters to emergency oxygen units on three space suits needed for Shuttle Atlantis' planned September supply mission (STS-106) to the International Space Station but repairs are not expected to delay launch. ["Spacesuit contaminants found," *The Orlando Sentinel*, July 15, 2000, p A-4.]

JUNE 15: Fifteen months after the first test flight was to have occurred, the troubled X-33 test launch vehicle still is on the ground in California with no flight date in immediate sight. The delay is the latest in a long string of delays since March 1999, when the X-33 was to have been sent aloft for the first time. They have been brought on by a range of technical problems, including recent difficulties with the X-33's lightweight fuel tanks. Once produced, the VentureStar is to have the capability of flying payloads into space quickly at far lower costs than current vehicles, promising much new business for Kennedy Space Center and other launch sites around the country. In addition to flying payloads, the VentureStar also is intended to possibly carry humans one day. But the X-33 delays mean that production of VentureStar, if it occurs, might be at least five years away. ["Troubled X-33 still not ready to fly," *Florida Today*, June 16, 2000, p 1A & 2A.]

JUNE 16: A team of NASA scientists has been formed to find out why a key piece of a shuttle engine failed a recent test. The engine piece, a modified turbopump, heated up quickly during a test firing of the engine June 16 at the Stennis Space Center in Mississippi, NASA officials said. ["NASA inquiry examines failed shuttle-engine test," *Florida Today*, June 25, 2000, p 1A.]

◆ As part of the KSC reorganization of functions and personnel the newly created Safety, Health and Independent Assessment Directorate (SH&IA) is now responsible for providing the single point-of-contact within KSC for all external assessments, audits and/or inspections. Accordingly, the OIG/GAO Liaison function has been transitioned from the Management Planning Division under the Administration Office, to the External Assessment Team under the Assessment Division within SH&IA. [Memoranda (NASA). Shannon D. Bartell. Subject: "Audit Liaison Representative and External Assessment Responsibility," June 16, 2000.]

JUNE 17: History is made almost every day at NASA's Kennedy Space Center, so it is appropriate that an artifact from the space program's earliest days is on display at the Visitor Complex. Gus Grissom's Mercury capsule will begin a three-year nation-wide tour starting at KSC. The exhibit, "The Lost Spacecraft: Liberty Bell 7 Recovered," will be open to the public until Sept. 17. The capsule, launched on July 21, 1961, sank during recovery operations in the Atlantic Ocean when the explosive hatch fired prematurely. While Grissom escaped unharmed, the capsule filled with water and sank in 16,043 feet of water.

A Discovery Channel expedition led by Curt Newport recovered the capsule last year. It was located approximately 300 miles east of Cape Canaveral. ["Rescuer recalls sinking of Liberty Bell capsule," *Florida Today*, June 14, 2000, p 1A & 3A. "Space artifact comes home," editorial, *Florida Today*, June 14, 2000, p 12A. *A Summary of Major NASA Launches, October 1, 1958 – December 31, 1989*, June 1992, p V-3.]

JUNE 19: An Indiana Democrat plans to ask fellow members of Congress to drop the Russians from the International Space Station project. U.S. Rep. Tim Roemer plans to attach his "drop Russia" amendment to NASA's $14 billion budget request this week, spokesman for the congressman's office said Monday. Roemer attempts to kill the space station project every year. ["Congressman wants to drop Russians from station project," *Florida Today*, June 20, 2000, p 1A & 2A.]

JUNE 20: In the Space Station Processing Facility (SSPF), workers last week attached an S-band Antenna Support Assembly (SASA) to the Integrated Truss Structure Z1, an element of the International Space Station. The SASA antenna is primarily for local communications between the orbiter and Space Station. The Z1 is the payload on mission STS-92, the fifth flight to the Space Station, scheduled this fall. ["Space Station hardware being readied for mid-fall mission, STS-92," *KSC Countdown*, June 20, 2000.]

◆ Ed Gormel, Executive Director of the Joint Performance Management Office, retired. ["Ed Gormel, head of JPMO, to retire," *KSC Countdown*, June 20, 2000.]

JUNE 21: The House easily defeated Wednesday yet another attempt to cancel funding for the $60 billion International Space Station. The annual bill, offered by Rep. Tim Roemer, D-Ind., was not even close: 98 yes votes to 325 against. ["Funding for space station survives," *Florida Today*, June 22, 2000, p 2A.]

JUNE 22: KSC Center Directory Roy Bridges and Commander of the 45th Space Wing Brig. Gen. Donald P. Pettitt have signed an interagency agreement establishing the Joint Planning and Customer Service (JPCS) office. The agreement is intended to serve as a "one-stop shop" for new customers of the two federal agencies and expand the Air Force/NASA partnership. The JPCS office is temporarily located at Patrick AFB, with Rick Blucker serving as director. ["KSC/AF Sign Interagency Agreement," *KSC Countdown*, June 27, 2000.]

JUNE 26: The long-delayed launch of the Russian Zvezda Service Module to the International Space Station has been scheduled for July 12, NASA and Russian officials announced. The window for the launch from the Baikonur Cosmodrome in the Republic of Kazakstan has not been set yet, but launch is expected around 1 a.m. EDT. Once in space, Zvezda will be mated with the only two pieces of the station in orbit: the Russian-built Zarya and the U.S.-built Unity. ["Launch date set," *Florida Today*, June 27, 2000, p 5A.]

◆ Two high-profile lawmakers are seeking an investigation into design difficulties and at least $200 million in projected cost overruns involving a planned U.S.-designed piece for the $60 billion International Space Station. The investigation is being requested by U.S. Sen. John McCain, R-Ariz., and U.S. Rep. James Sensenbrenner, R-Wis. The U.S. piece in question, a propulsion module, is being designed by The Boeing Co. of Seattle.

Construction has not started. Under contract with NASA, the module was to be used if the Zvezda Russian service module was not launched. The Zvezda, which has been delayed two years, is scheduled for launch July 12, but NASA plans to go ahead with the propulsion module, using it to guide the station as a backup to Russian equipment. ["Lawmakers want investigation of space station costs," *Florida Today*, June 27, 2000, p 1A & 5A.]

JUNE 27: Determined to avoid a repeat of last month's problem with boaters, the Air Force has taken several steps to make sure a launch debris danger zone off Port Canaveral is kept clear for a planned rocket launch scheduled for Friday (June 30). Among the Air Force's steps to keep the zone clear include: The creation of a telephone hot line with up-to-the-minute launch information; a series of pre-launch meetings with boaters at Port Canaveral; distribution of a danger-zone map through newspapers, fliers, Web sites and TV stations; and increase in Air Force and Coast Guard surveillance throughout the area before launch. The launch impact danger zone extends 14 miles north, 6 miles south and 80 miles out into the Atlantic from launch pad 36B. ["Air Force works to clear launch zone," *Florida Today*, June 28, 2000, p 1B.]

JUNE 28: NASA and industry engineers are working to resolve a technical issue involving an Atlas 2A rocket, which had been scheduled to launch today with a NASA tracking and communications satellite aboard. To give engineers more time to work on the problem, the launch from Cape Canaveral Air Force Station has been rescheduled to Friday (June 30). ["Engineers work for Friday launch of Atlas 2A rocket," *Florida Today*, June 29, 2000, p 1B.]

◆ In the SSPF, Boeing engineers removed the "bay window" on the U.S. Laboratory Destiny to clean and replace a secondary seal observed during testing to have a small pressure leak. Destiny was moved in a canister for a transfer to the O&C Building. It is to be installed into an altitude chamber this week for a pressure test of the laboratory itself. A component of the International Space Station, Destiny is scheduled to fly on mission STS-98 in early 2001. ["U.S. Lab to undergo pressure test in O&C altitude chamber," *KSC Countdown*, June 29, 2000.]

JUNE 30: An Atlas 2A rocket successfully lifted off from Cape Canaveral Air Force Station at 8:56 a.m. carrying a Tracking and Data Relay Satellite. TDRS-H joins six other satellites that link astronauts to Earth, as well as other orbiting satellites to their ground stations. ["Atlas 2 liftoff a success," *Florida Today*, July 1, 2000, p 1A & 3A.]

◆ Despite a new launch awareness effort, the Air Force and Coast Guard still had to cope with two boats that wandered into a danger zone during the launch attempt of an Atlas 2A. The rocket, with a NASA satellite aboard, was launched from Cape Canaveral Air Force Station at 8:56 a.m., halfway into a 40-minute launch window. By the time the boat was shooed out of the danger zone, 18 minutes of the window had been lost. A shrimp boat also strayed into the zone. In that case, a Coast Guard vessel carried the boat's three occupants out of the zone, and their craft was left behind. ["2 boats stray into danger zone," *Florida Today*, July 1, 2000, p 3A.]

DURING JUNE: Edward J. Heffernan has been named Chief of Staff at NASA Headquarters by Administrator Daniel S. Goldin. ["Heffernan named NASA chief of staff," *The Brevard Technical Journal*, June 2000, p 2.]

◆ Kennedy Space Center is now in negotiations with Gainesville's FLAD Inc. to design the proposed Space Experiment Research and Processing Laboratory (SERPL) facility, and with Orlando's ZHA Inc. to support the Cape Canaveral Spaceport master planning for KSC and the Cape Canaveral Air Force Station. This NASA master planning contract and effort will be coordinated with the Spaceport Authority's ongoing five-year transportation plan. ["NASA selects contractors for SERPL and planning," *The Brevard Technical Journal*, June 2000, p 30.]

JULY

JULY 4: The final hurdle to restarting construction of the International Space Station was cleared when a modified Russian Proton rocket flew successfully, clearing the way for the launch of Zvezda. ["Russians to launch Zvezda next week," *Florida Today*, July 6, 2000, p 1A.]

JULY 6: Dave Weldon, U. S. Rep., R-Palm Bay, and Patsy Ann Kurth, state Sen., D-Malabar, vowed to fight for more funding and workers for Kennedy Space Center, and for funding for research facilities at KSC. They both support manned missions to Mars and will continue to support the International Space Station. In short, the space program has an advocate on Capitol Hill – regardless of who wins in November. ["Weldon, Kurth vow support for space," *Florida Today*, July 7, p 1B & 2B.]

◆   Atlantis, in the Orbiter Processing Facility (OPF) bay 3, is scheduled for engine installation this week in preparation for mission STS-106, while Discovery is in the Orbiter Processing Facility (OPF) bay 1 waiting for the arrival of a Space Station component. Launch date for Discovery, for mission STS-92, has been moved to October 5 at approximately 9:49 p.m. ["Atlantis gets new engine; Discovery launch moved to Oct. 5," *KSC Countdown*, July 6, 2000.]

◆   Pressurized mating adapter –3 is installed in Discovery (STS-92), in the Orbiter Processing Facility [(PF) bay 1. ["Orbiters in OPF undergoing system tests and checks," *KSC Countdown*, July 20, 2000.]

JULY 7: E'Prime Aerospace Inc. is edging closer to an agreement with Kennedy Space Center that could allow it to launch rockets there. The Commercial Launch Vehicle Programs Agreement would not be a contract to allow E-Prime to use KSC land or launch support services but would set in motion a series of studies that could pave the way for a more concrete agreement. ["Rocket company, KSC closer to a deal," *Florida Today*, July 7, 2000, p 10C & 9C.]

JULY 12: A Russian Proton rocket was launched at about 12:56 a.m. from Baikonur Cosmodrome in Kazakstan carrying the Zvezda Service Module. The launch came after two years of delays getting the component off the ground. The delays were caused by previous Russian financial problems. The Zvezda Service Module is to join two sections of the International Space Station now in orbit. It will provide living quarters and power and steering to keep the station in a safe orbit about 240 miles above Earth. ["Russians send Zvezda into orbit," *Florida Today*, July 12, 2000, p 1A.]

JULY 14: The Lockheed Martin Corporation's Atlas 2AS successfully launched a high-powered TV satellite, Echostar 6, for the EchoStar Communications Corp., Littleton, Colo., from Cape Canaveral Air Force Station, launch pad 36B, at 1:21 a.m. EchoStar, through its DISH Network has the capacity to offer more than 500 satellite TV channels throughout the U. S. in a battle against cable TV companies. The satellite was placed in an orbit about 22,300 miles above Earth ["Air Force ready for Delta 2 launch," *Florida Today*, July 15, 2000, p 3A.]

JULY 15: A black-tie gala commemorating the July 24, 1950, launch of the first rocket from the Cape, Bumper 8, was held at the Radisson Resort in Port Canaveral. Guest speakers included U.S. Rep. Dave Weldon and Kennedy Space Center (KSC) Director Roy Bridges, and sponsored by the Cape Canaveral Chapter of the Air Force Association. ["Event marks first launch from Canaveral," *Florida Today*, July 16, 2000, p 2A.]

◆   Today marks the 25th anniversary of the Apollo Soyuz Test Project (ASTP) which was the first cooperative human space flight endeavor. The nine-day mission began July 15, 1975 with the launch of a Saturn IB launch vehicle from Launch Complex 39B. ["Anniversary of Apollo Soyuz Test Project observed," *NASA Note to Editors #N00-30*, July 11, 2000.]

JULY 16: Boeing's Delta 2 successfully launched from Cape Canaveral on pad 17A at 5:17 a.m. carrying a Global Positioning System (GPS) satellite built for the Air Force by Lockheed Martin. There were no technical issues, weather problems, or boaters encroaching on a launch-debris danger zone. ["Boeing to give Delta 3 a third try in August," *Florida Today*, July 17, 2000, p 1A.]

JULY 17: A major shuffling of duties of the safety officers in the space program has some critics worried about civilian safety during launches. The safety officers, a combination of civilian and Air Force engineers, certify that rockets and ground equipment are ready for launch and press the destruct button if missions veer off course. The Air Force's Materials Command is taking over the job of making sure rockets are built safely and ready to fly and safety officers at the Cape and Vandenberg are being moved under the authority of operations groups, which are primarily responsible for getting missions off the ground. ["Safety issues tough off debate," *The Orlando Sentinel*, July 17, 2000, p A-4]

JULY 18: A ribbon-cutting ceremony will be held at 10 a.m. for its latest new addition, the Vapor Containment Facility (VCF) which is located next to the Space Station Processing Facility. The facility contains ammonia servicing equipment, used in connection with flight hardware for the International Space Station. The ceremony will feature Center Director Roy Bridges and remarks by International Space Station Program Deputy Manager Jay H. Greene along with other featured speakers. ["New KSC facility opens with ribbon cutting July 18," *KSC Countdown*, July 13, 2000.]

◆   The VAB is nearing the end of modifications that will make it a safe haven for NASA's orbiters if a hurricane threatens the area. To give the fleet more safety, NASA is making room in two spare high bays that have been used to store external fuel tanks for the shuttles. ["Upgrades to VAB will help protect shuttles from storms," *Florida Today*, July 18, 2000, p 8A.]

◆   The Space Station Remote Manipulator System (SSRMS), also known as the "Canadian arm," was moved to a palette in the Space Station Processing Facility for testing. The SSRMS is an essential component of the International Space Station and will play a key role in assembly and maintenance. The SSRMS will be on Shuttle mission STS-100, flight 6A, to the Space Station. ["Canadian robotic arm on the move in SSPF for testing", *KSC Countdown*, July 18, 2000.]

◆ The first train derailment in KSC history occurred while the train, hauling six booster segments from the storage yard to the processing facility, was moving at 3 m.p.h. (4.8 k.p.h.). It occurred near Gate 2C north of the KSC Industrial Area. The six cars containing the segments were each separated by a water filled spacer car. The back wheels of one car and the front wheels of another car slipped off the tracks as the engine attempted to connect them with the other three train cars. No one was injured and none of the segments was damaged, but about 30 feet of track sustained moderate damage. An investigation board was set up to probe the accident and could release its findings in a couple of weeks. [Web posted. (2000). Train hauling shuttle booster segments derailed at Kennedy Space Center [Online]. Available WWW: http://www.space.com/ [2000, July 19]. "Train derailment unlikely to delay February launch," _Florida Today_, July 22, 2000, p 1B. "Final word on derailment could be out next month," _Florida Today_, July 30, 2000, p 1A.]

JULY 20: Flight control testing is underway on the maneuvering system and reaction control system on orbiter Atlantis (STS-106) in the Orbiter Processing Facility (OPF) bay 3. The orbiter maneuvering system pod checks are being performed on Discovery in the OPF bay 1 in preparation for mission STS-92. Endeavour, inside Orbiter Processing Facility (OPF) bay 2, is having docking system pyrotechnic system checks before the STS-97 launch date. ["Orbiters in OPF undergoing system tests and checks," _KSC Countdown_, July 20, 2000.]

JULY 22: Work on about 30 feet of railroad track torn up in the July 18th train derailment is expected to begin today. The segments, which form the rocket boosters that power the space shuttles, are currently in the derailed train and were being transported to a nearby processing facility. These segments will be used in a February mission aboard Discovery (STS-102) so the accident will unlikely cause any launch delays. Among the possible causes of the accident are a switch that shifts the trains between tracks, or a weak spot on the tracks. ["Train derailment unlikely to delay February launch," _Florida Today_, July 22, 2000, p 1B.]

◆ The Air Force Space and Missile Museum on Cape Canaveral Air Force Station will be open to the public July 22-24, free of charge, as part of the yearlong celebration commemorating the 50th anniversary of the Bumper 8 launch. ["Visitors can drive to Cape, view space museum for free," _Florida Today_, July 22, 2000, p 1A.]

◆ The two derailed cars from the July 18th railroad accident at KSC were placed back on the railroad tracks at 8:45 a.m. ["Final word on derailment could be out next month," _Florida Today_, July 30, 2000, p 1A.]

JULY 23: Workers began repairing the 30 feet of damaged railroad track. That project was delayed due to recent afternoon thunderstorms. ["Final word on derailment could be out next month," _Florida Today_, July 30, 2000, p 1A.]

◆ Twenty-one high school students from Brevard and four surrounding counties are working at Kennedy Space Center (KSC) as part of NASA's 20th annual Summer High Apprenticeship Research Program. The goal of the program is to get underrepresented students interested in careers in engineering and science. It is an eight week paid apprenticeship in which the students are teamed with NASA mentors. ["Students work with NASA mentors," *Florida Today*, July 23, 2000, p 1A & 3A.]

JULY 24: Today marks the 50th anniversary of the Bumper 8 launch from Pad 3 on Cape Canaveral. A ceremony was held at Launch Complex 26, which included the launching of model rockets and speeches. Among those in attendance were Roy Bridges, KSC Center Director; Jimmy Morrell; Winston "Bud" Gardner; State Sen. George Kirkpatrick; Brig. Gen. Don Pettit; and Norris Gray. ["Ceremony commemorates Cape's 1st rocket launch," *Florida Today*, July 25, 2000, p 1B & 2B. "Bumper 50th anniversary ceremony held at CCAFS," *KSC Countdown*, July 27, 2000.]

◆ State Sen. George Kirkpatrick from Gainesville, was honored by the Kennedy Space Center and the Florida Space Business Roundtable during a ceremony commemorating the 50th anniversary of the first rocket launch from the Cape for his support of space issues. KSC Director Roy Bridges said, "(He) has shown enthusiasm toward the space program and its long-term impact on the economy of the state of Florida." ["Senator a space fan," *Florida Today*, August 27, 2000, p 1E.]

◆ A monthly meeting of the National Space Club Florida Committee attracted nearly 225 guests to salute the original Bumper Launch Team. It was a luncheon held at the Kurt H. Debus Conference Center at KSC Visitor Complex. Among those who were recognized for their involvement in the 1950 launch were Dr. William Pickering, Homer Joe Stewart, Robert Droz, Herman Banks, Hiroyuki Hashimoto, and Al Richardson, all of California; Norris Gray, Dick Jones, Al Siple, E. M. Bain, Ed Belcher, and Lloyd Behrendt, all of Brevard County; Howard Hoge of Maryland; and Konrad Dannenberg, of Alabama. ["Space Club fete draws multitude," *Florida Today*, July 29, 2000, p 1D.]

JULY 25: The STS-106 crew spent several days last week taking part in Crew Equipment Interface Test activities, primarily at SPACEHAB, Cape Canaveral, Florida. The crew consists of Commander Terrence W. Wilcutt, Pilot Scott D. Altman, and Mission Specialists Edward T. Lu, Yuri I. Malenchenko, Boris V. Morukov, Richard A. Mastracchio, and Daniel C. Burbank. ["Atlantis crew prepares for STS-106 launch just six weeks away," *KSC Countdown*, July 25, 2000.]

◆ Experts in the fields of biology, chemistry, physics and general science and engineering will discuss near- and long-term research on the International Space Station today at 3:30 p.m. EDT at the KSC press site auditorium. Participants will include Dr. Julie Swain, Dr. Mary Musgrave, Dr. Milburn Jessup, Dr. Kathy Clark, and Dr. Ron Sega. ["ISS press conference," *KSC Countdown*, July 25, 2000.]

◆ Russia's Zvezda Service Module docked flawlessly at the International Space Station at 8:45 p.m. EDT, two weeks after its launch. ["Zvezda docks with station," *Florida Today*, July 26, 2000, p 1A & 2A.]

JULY 26: Seventeen men and women have been selected for the astronaut candidate class of 2000. Nicole Stott, formerly an engineer at KSC working for Tip Talone, director of the Space Station Hardware Integration Office, was one of those chosen. [NASA Names

Astronaut Candidate Class of 2000," NASA News Release #00-116, July 26, 2000. "Former KSC engineer may fly in space some day," _Florida Today_, September 3, 2000, p 1A & 2A.]

◆   The truss segment of the International Space Station called "P1" arrived at KSC aboard the "Super Guppy" transport aircraft. It will be attached to the port side of the center truss, S0, and will be the third assembled to the ISS. It's due to be launched in the Spring of 2002. ["P-1 truss segment arrives at KSC," _KSC Countdown_, July 27, 2000.]

JULY 28: The Astronaut Crew Quarters opened its doors to visiting children during "Take our children to work day" . ["The way the astronauts live," _KSC Countdown_, August 1, 2000.]

JULY 30: The on-site investigation into the July 18 railroad accident at KSC is over. But the advisory board studying the accident probably won't release its findings and recommendations until August. ["Final word on derailment could be out next month," _Florida Today_, July 30, 2000, p 1A.]

◆   An International Space Station truss segment, P4, arrived at the Space Station Processing Facility at KSC. It arrived via trailer from Tulsa, OK, where it was manufactured and assembled by The Boeing Co. ["P4 truss arrives at SSPF via trailer from Tulsa," _KSC Countdown_, August 3, 2000.]

JULY 31: The Zenith 1 (Z-1) Truss was inspected in the Space Station Processing Facility and OK'd by some of the shuttle Discovery crew that is to deliver it to the International Space Station on the STS-92 mission in October. Altogether, there are about 250,000 pounds of station parts waiting at Kennedy Space Center (KSC) for delivery into space. ["Astronauts inspect station's 'backbone'," _Florida Today_, August 2, 2000, p 1A.]

◆   The Zenith 1 (Z-1) Truss, the cornerstone truss for the International Space Station, was symbolically transferred from The Boeing Co. to NASA. The transfer signals the completion of prelaunch processing at KSC's Space Station Processing Facility. The Z-1 is scheduled to fly in shuttle Discovery's payload bay on STS-92 targeted for launch October 5. ["Z-1 truss transferred to NASA at ceremony," _KSC Countdown_, August 1, 2000.]

DURING JULY: The state of Florida has partnered with NASA to develop a research facility and commerce park at KSC. A $14 million investment included in the FY 2001 budget recently signed by Governor Jeb Bush will support the construction of a Space Experiment, Research & Processing Laboratory (SERPL) at KSC, as well as a roadway to serve the SERPL and a 400-acre Space Station Commerce Park on KSC property. It will be an approximately $43 million facility of which NASA is contributing $9 million for design and activation, Delaware North Park Services is allocating $4 million for the roadway. Funds for completion will be proposed for the state's FY-2002 budget. The planned 100,000 square-foot facility will be constructed by the state over a 25-month period and will feature shared use laboratories. Upon completion, the SERPL facility will be co-managed by NASA and the Florida Space Research Institute, using the University of Florida as its lead institution. ["State partners with NASA to develop research facility and commerce park at KSC," _Brevard Technical Journal_, July 2000, p 3 & 9.]

◆   The Titusville-based Command and Control Technologies Corporation (CCT) recently outgrew its 1,000 square-foot facility in the Florida/NASA Business Incubation Center at the Brevard Community College in Titusville. The company will move into a 4,100 square-foot space at 1425 Chaffee Drive in Titusville. A large portion of CCT's revenue comes from the sale and installation of launch control systems and spaceport consulting. ["CCT outgrows incubation center," _Brevard Technical Journal_, July 2000, p 9.]

AUGUST

AUGUST 1: A bag containing urine was found at on the 14th floor of Launch Complex 40 near a work area that requires strict sanitary conditions to protect equipment. A Titan 4 rocket is on the pad. About 20 people, all workers at the Titan launch site, are being investigated by Lockheed Martin Corp., with assistance from the Air Force's Office of Special Investigations. None of the foreign material came into contact with the rocket or the satellite and no one's health was affected. ["Bag of urine found on Cape launch pad," *Florida Today*, August 5, 2000, p 1A & 2A.]

AUGUST 3: The annual Community Leaders Briefing is being held this morning at the KSC Visitor Complex. Center Director Roy D. Bridges and James L. Jennings, Deputy Director, will meet with various community leaders about the long term viability of KSC and benefits the space program contributes to the community. The theme is "Milestones of the Millennium" and the leaders will hear about KSC's vision, budget, employment trends, launch outlook, and future goals. ["Community leaders breakfast," *KSC Countdown*, August 3, 2000.]

AUGUST 4: The motion picture "Space Cowboys" premiered locally at the Oaks 10 Theater in Melbourne. A tale of four aging ex-pilots who finally see their decade-old dream of going into space come true was filmed, in part, at Kennedy Space Center (KSC) last year. Scenes filmed at KSC included some on the shuttle launch pad and some of a shuttle being launched. ["Residents pack Melbourne movie theater for local premiere," *Florida Today*, August 5, 2000, p 1B.]

◆ A group of twenty-three librarians end a week of learning about space at the Visitor Complex, courtesy of NASA, with launches of model rockets they built during the week. This is the first group of librarians offered the space education courses, according to Steve Dutczak of the NASA Education Office. The librarians were selected by the directors of Brevard, Volusia, and Orange Counties. Another librarian group will be attending this space education opportunity next week. ["Librarians' rockets take flight from KSC," *Florida Today*, August 6, 2000, p 1B & 2B.]

AUGUST 6: A Russian supply spacecraft, the unmanned Progress on a Soyuz rocket, launched successfully at 2:26 p.m. EDT, carrying fuel, supplies, and equipment for the International Space Station. ["Russian supply craft lifts off," *Florida Today*, August 7, 2000, p 1A.]

◆ The 2000 International Law Enforcement Games opened today at the KSC Visitors Complex. More than 1,850 law enforcement officers and their families were present for the opening ceremonies. Participants from 15 countries and 37 states gathered at the Rocket Garden for parades and torch lighting. Brevard County Sheriff Phil Williams, Brevard County Commissioner Helen Voltz, and astronaut Sam Durrance provided opening comments. ["KSC hosts opening ceremonies for law enforcement games," *KSC Countdown*, August 8, 2000.]

AUGUST 7: Space Shuttle Atlantis moved from its hangar in the Orbiter Processing Facility (OPF) bay 3 to the Vehicle Assembly Building where it will be attached to an external fuel tank and solid rocket boosters. Atlantis is scheduled to rollout to Launch Complex 39B on August 13th where preparations for mission STS-106 will begin. ["Atlantis scheduled for Sept. 8 flight to International Space Station," *Florida Today*, August 8, 2000, p 1B.]

AUGUST 8: A 20-acre brush fire on KSC, which came within 80 yards of several fueled booster-rocket segments was brought under control. The segments were inside train cars which had derailed July 18. The fire had started from a lightning strike August 5, and subsided but had re-ignited. ["Firefighters contain 20-acre blaze at KSC," *Florida Today*, August 9, 2000, p 1A.]

◆   The unmanned Russian supply spacecraft, Progress, docked with the International Space Station at 4:13 p.m. Progress is carrying about 1,300 pounds of food, supplies and equipment that will be used by the resident crews. ["Russian supply vehicle docks with space station," *Florida Today*, August 9, 2000, p 1A & 2A.]

◆   The 21st annual National Aerospace FOD Prevention Conference opened in Orlando. Foreign Object Debris is a safety challenge for the aviation/aerospace industry, especially at Kennedy Space Center (KSC) facilities. Speakers included astronaut Kay Hire and United Space Alliance Deputy Space Shuttle Program Manager Loren Shriver. Topics at the conference covered innovative programs and processes being used by the industry to prevent FOD. ["Speaking about FOD," *KSC Countdown*, August 10, 2000.]

AUGUST 12: Space Shuttle Atlantis rolled into the Vehicle Assembly Building's newly modified high bay 2 on the west side for the final Safe Haven fit check. Atlantis was then rolled out to Launch Pad 39B in preparation for mission STS-106. Atlantis is scheduled to launch September 8 at 8:31 a. m. EDT. ["Shuttle Atlantis okays Safe Haven fit, then moves to Pad 39B," *KSC Countdown*, August 15, 2000.]

◆   Today marks the 40th anniversary of the first successful Delta launch which occurred August 12, 1960. The Delta launch vehicle successfully orbited Echo 1. [Delta is up, milestone was missed [Online]. Available WWW: http://www.floridatoday.com/ [2000, August 23].]

AUGUST 15: The crew of Space Shuttle Atlantis, STS-106, arrived at KSC about 7 p.m, lead by commander Terrence W. Wilcutt and crew members Scott D. Altman, pilot; mission specialists Daniel C. Burbank, Edward T. Lu, and Richard A. Mastracchio, plus Yuri Ivanovich Malenchenko and Boris V. Morukov from Russia. The Atlantis crew will spend the next three days conducting a launch dress rehearsal. ["Atlantis crew arrives," *Florida Today*, August 16, 2000, p 1B.]

◆   The KSC Visitor Complex has stopped selling space shuttle launch tickets separately, and instead is selling them as part of higher priced packages to other Space Center Attractions. The new ticket structure is intended to encourage visitors to learn more about NASA's space program by seeing all of the Space Center's attractions, said Don LeBlanc, spokesman for Delaware North Parks Services of Spaceport, Inc. ["KSC halts launch-only ticket sales," *Florida Today*, August 16, 2000, p 1A & 2A.]

AUGUST 16: The STS-106 crew began Terminal Countdown Demonstration Test (TCDT) activities. They took turns driving the M113 armored personnel carriers and practiced fire suppression training. Thursday, August 18, they will practice emergency egress from the orbiter at Launch Pad 39B, and Friday, August 19, is the dress rehearsal for launch with a simulated countdown. ["STS-106 crew get dress rehearsal for launch in 3 weeks," *KSC Countdown*, August 17, 2000.]

◆ A Canadian-built mechanical arm (Space Station Remote Manipulator System or SSRMS) was officially turned over to NASA for final processing to the International Space Station. It's scheduled for launch on STS-100 in April 2001. ["Canada's contribution ready to go to station," *Florida Today*, August 17, 2000, p 1A. "Canadian arm moves one step closer to destination: Space Station," *KSC Countdown*, August 17, 2000.]

AUGUST 17: Shuttle Atlantis, STS-106, crew met with new media at KSC, in front of Launch Pad 39-B. The mission is to prepare the orbiting outpost for the arrival of the first resident crew in November. ["Atlantis crew, spacesuits prepared for Sept. 8 launch," *Florida Today*, August 18, 2000, p 1B.]

AUGUST 18: David A. King, Director of Shuttle Processing, announced the creation of a new position in the Shuttle Processing Directorate of KSC: the Launch Manager. The first two Launch Managers are John A Guidi and Edward J. Mango. They are the single point of contact in the Directorate for all Shuttle mission flow-related issues. They also serve, on an alternating basis, as Assistant Launch Directors for assigned missions and represent the Launch Director when required. Mr. Guidi is assigned Shuttle Processing mission responsibilities for STS-92, STS-98, and STS-100 and Mr. Mango for STS-97, STS-102, and STS-104. Thereafter, they will alternate mission responsibilities. [Memoranda (NASA). David A. King. Subject: "Launch Manager, Shuttle Processing Directorate," August 18, 2000.]

AUGUST 20: Today marks the 25th anniversary of the Titan Centaur 4 launch carrying the Viking 1 spacecraft, which landed on Mars July 20, 1976. [Delta is up, milestone was missed [Online]. Available WWW: http://www.floridatoday.com/ [2000, August 23].]

AUGUST 21: With 250 shuttle program workers watching, Space Shuttle Discovery was moved from the Orbiter Processing Facility (OPF) bay 1 to the Orbiter Processing Facility (OPF) bay 3 as flight preparation work continues. Discovery's launch on mission STS-92 will mark the 100th time a shuttle has flown from Kennedy Space Center. ["Shuttle set for 100th flight from KSC." *Florida Today*, August 22, 2000, p 3A. "Discovery rolls from OPF bay 1 to bay 3," *KSC Countdown*, August 22, 2000.]

AUGUST 22: Charles Starowesky, a former Navy sailor, pleaded guilty in federal court to possessing stolen government property. He retrieved a heat-shield tile from the Space Shuttle Challenger accident site and tried to sell it over the Internet. He had held on to the tile for 14 years in violation of federal law. Starowesky had tried to sell the heat-shield tile on the Internet site eBay on October 28, 1999 according to a statement from NASA's Office of Inspector General. He was sentenced to two years probation on a count of possessing government property with intent to sell. [Web posted. (2000) Man sentenced for attempted sale of Challenger debris [Online]. Available WWW: http://www.space.com/ [2000, August 23].]

AUGUST 23: The Boeing Company's Delta 3 rocket successfully carried a simulated satellite into space at 7:05 a.m. The payload is a 9,586 pound steel spool built to simulate a real satellite's size and weight. The test was designated DM-F3. Boeing has added stripes, triangles and reflectors to the spool so that Air Force and University of Colorado scientists can use it to fine tune tracking systems and conduct other studies during the payload's estimated lifetime of 15 years. ["Boeing's Delta 3 rocket finally flies with success," *Florida Today*, August 24, 2000, p 1A & 7A. "Boeing seeks 1st success of Delta III," *The Orlando Sentinel*, August 23, 2000, p A-6. "Delta III succeeds, but will it sell?", *Aviation Week & Space Technology*, August 28, 2000, p 48-49.]

◆ After a meeting, KSC managers decided that Hurricane Debby is not yet a serious enough threat to warrant a move of Space Shuttle Atlantis from Launch Pad 39B to the Vehicle Assemble Building. ["Managers leave shuttle Atlantis on launch pad," *Florida Today*, August 24, 2000, p 3A.]

◆ Today, prelaunch propellant loading (for STS-106) of Atlantis's storage tanks occurred. The payload bay doors are scheduled to be closed for flight August 30. Launch is targeted for September 8 at 8:45 a.m. ["Atlantis, Discovery on track for respective launches," *KSC Countdown*, August 24, 2000.]

AUGUST 24: Shuttles Atlantis (STS-106) and Discovery (STS-92) are on target to keep their upcoming launch dates. Discovery was moved from the Orbiter Processing Facility (OPF) bay 3 to the Vehicle Assembly Building (VAB) for mating to the external tank and the solid rocket boosters. The move was delayed 3.5 hours while shuttle managers reviewed weather and technical reports. Atlantis remains on Launch Pad 39B. Discovery is to head to Launch Pad 39A next week providing the rare event of two shuttles on Kennedy Space Center's launch pads at the same time. ["Next two shuttle flights on schedule," *Florida Today*, August 25, 2000, p 1A & 2A. "Atlantis, Discovery on track for respective launches," *KSC Countdown*, August 24, 2000.]

◆ Retired Air Force Maj. Gen. Michael Butchko has been named the new program manager for Space Gateway Support (SGS), which provides day-to-day services to Kennedy Space Center and Cape Canaveral Air Force Station. He will begin his new position October 1 and will replace William Hickman program manager since 1998. Hickman has been assigned another position with Logicon Inc., a Northrop Grumman Corp. subsidiary. ["Space Gateway has new leader," *Florida Today*, August 25, 2000, p 1C.]

AUGUST 25: Charles Williams, a Boeing Co. official, announced Shuttle Columbia has five times the problems that NASA's three other shuttles showed during their wiring inspections. The inspections showed 600 to 700 wiring defects in Atlantis, Discovery and Endeavour, but Columbia showed about 3,500. Despite Columbia's problems, KSC spokesman Bruce Buckingham said that NASA's flight schedule should not be affected as the agency has no plans to use Columbia for station missions and the other three Shuttles have had their wiring problems corrected. ["3,500 wiring problems found on Columbia," *Florida Today*, August 26, 2000, p 1A & 2A.]

◆ Orbiter Discovery (STS-92) was lifted into High Bay 3 of the Vehicle Assembly Building (VAB) and mated to the external tank and solid rocket boosters on the Mobile Launcher Platform. The Flight Readiness Review for Atlantis' mission, STS-106, is scheduled for August 29. ["10 days to targeted launch date for STS-106 mission," *KSC Countdown*, August 29, 2000.]

AUGUST 28: As a safety precaution during hurricane season, NASA managers will now allow two shuttles to be positioned on KSC launch pads at the same time. Kennedy Space Center spokesman Joel Wells said the decision does not mark a new one-shuttle-only pad policy for NASA and simply is a precautionary measure. Discovery will remain inside the Vehicle Assembly Building (VAB) until September 11, three days after Atlantis (STS-106) is to fly. ["Discovery won't roll until Atlantis lifts off," *Florida Today*, August 29, 2000, p 1A.]

◆ A silver handle from the spacecraft Apollo 11 has been sold at auction by the auction house Butterfield & Butterfield of San Francisco for $34,500. The sale is conditional pending the results of a NASA inquiry concerning ownership of the item. The handle has been in the possession of Charles Brown, a former radiation safety officer at Kennedy Space Center who performed test on the silver handle in the 1970's. Brown kept the handle in a safe after leaving NASA and performed periodic tests on it for the administration to check for radiation exposure. The Office of Inspector General for NASA said that an ongoing investigation is trying to determine ownership. [Web posted. (2000). NASA could nullify auction of Apollo 11 handle [Online]. Available WWW: http://www.cnn.com/ [2000, August 28].]

AUGUST 29: Jan Heuser, NASA's program manager for the new Space Experiment Research and Processing Laboratory, announced construction work is expected to begin by mid-2001 and should be completed around Spring 2003. The research laboratory will form the heart of a 400 acre commerce park at Kennedy Space Center. The project is an effort by Florida officials, with support from NASA, to expand KSC's historically significant launch role to include research activity. ["KSC lab to break ground mid-2001," *Florida Today*, August 30, 2000, p 1A.]

◆ A General Accounting Office [GAO] report was released stating NASA's space shuttle work force is understaffed and overstressed. Many key areas are not sufficiently staffed by qualified workers. The shuttles' work force is aging. It also noted that NASA has planned $2.2 billion worth of safety upgrades to the shuttle and is increasing the number of employees. It is also doing a better job of monitoring the health of its employees. ["Smaller staff seen as shuttle danger," *The Orlando Sentinel*, August 30, 2000, p A-7.]

AUGUST 31: NASA officials are satisfied that the spacesuits to be used on a spacewalk during the 11 day STS-106 mission are safe. The safety of the spacewalking suits sparked concerns this summer after NASA disclosed that an oily residue on the suits could have started a fire. The residue, which was found on emergency oxygen systems attached to the suits, was cleaned from three of the suits to prepare for this flight. The three suits have been loaded onto Atlantis. ["NASA deems spacewalking spacesuits safe," *Florida Today*, September 1, 2000, p 1A.]

SEPTEMBER 4: The Shuttle Atlantis' crew (STS-106) arrived at Kennedy Space Center around 8 p.m., encountering some fierce lightning while landing. The Atlantis crew will go through a range of activities to prepare them for the launch to the International Space Station. Launch is scheduled between 8:45 and 8:47 a.m. EDT, Friday, September 8. ["Lightning ushers in shuttle crew," _Florida Today_, September 5, 2000, p 1B.]

◆ A lightning strike is suspected of damaging a ground navigation system, called TACAN, located at the Shuttle Landing Facility. The system sends crucial distance information to the shuttle as it makes its landing. ["Bad weather may delay launch," _Florida Today_, September 6, 2000, p 1B.]

SEPTEMBER 5: The countdown to the STS-106 Atlantis launch, which began at 11:00 a.m., has proceeded without problems. ["Bad weather may delay launch," _Florida Today_, September 6, 2000, p 1B.]

◆ Lightning struck the Atlantis launch pad but caused no damage. The strike was the second at KSC in two days to hit equipment needed for Friday's Atlantis (STS-106) launch. Shuttle and launch equipment at the pad was protected by a 100-foot-tall lightning mast that the brunt of the bolt. The equipment, TACAN, damaged Monday was repaired, Shuttle Test Director Steve Altemus said Wednesday. ["Lightning strikes pad, but equipment is safe," _Florida Today_, September 7, 2000, p 8A.]

SEPTEMBER 6: The Checkout and Launch Control System (CLCS) at the Hypergolic Maintenance Facility (HMF) was declared operational in a ribbon cutting ceremony. The new control room will be used to process the Orbital Maneuvering System pods and Forward Reaction Control System modules at the HMF. ["CLCS ribbon cutting," _KSC Countdown_, September 7, 2000.]

◆ Engineers completed operations to load orbiter Atlantis' onboard cryogenic tanks for STS-106. No significant issues are being worked by the launch team at this time. [Bruce Buckingham. (2000) Kennedy Space Center Shuttle Processing Report [Online]. Available E-mail: domo@news.ksc.nasa.gov/subscribe shuttle-status [2000, September 7].]

SEPTEMBER 8: Space Shuttle Atlantis, STS-106, was launched at 8:45 a.m. EDT despite a 40% chance of weather delays. The launch window was only 2.5 minutes in order to save fuel for a possible 12th day added to the mission for outfitting the International Space Station. A decision to add an extra day will be made later in the mission. STS-106 is the second Atlantis mission to the International Space Station in four months. ["Shuttle on way to station," _Florida Today_, September 9, 2000, p 1A & 3A.]

◆ Lt. Cmdr. Dan Burbank, one of the seven astronauts aboard STS-106, is a Coast Guard officer, the second "Coastie" to fly into space. U.S. Transportation Secretary Rodney Slater attended the launch along with Coast Guard commandant, Adm. James Loy; Rear Adm. Thad Allen, the Seventh Coast Guard District commander; and Bruce Melnick, the only other Coast Guard astronaut and new a top executive the Boeing Co. ["Coast Guardsmen watch one of own fly into space," *Florida Today*, September 9, 2000, p 1A & 3A.]

SEPTEMBER 11: A $27 million Delta 4 rocket processing plant called the Horizontal Integration Facility was dedicated during a ceremony at Cape Canaveral Air Force Station. The seven-story 100,000 square-foot plant was funded through the Spaceport Florida Authority, which is developing the commercial space business at the Cape. The Boeing Co. plant will process rockets as part of an Air Force program called the Evolved Expendable Launch Vehicle (EELV) program. The Delta 4 as well as Lockheed Martin Corp.'s Atlas 5, are the cornerstone rockets in the EELV program. Both are the largest rockets either company ever has produced to deliver satellites into space. ["Boeing dedicates Delta 4 plant," *Florida Today*, September 12, 2000, p 1A & 3A.]

◆   Space Shuttle Discovery was moved from the VAB to Launch Pad 39A beginning about 2 a.m. The mission (STS-92), scheduled for an October 5 launch, will be the 100th in Shuttle history. The Zenith, or Z-1, payload was placed in a payload canister to prepare it for the flight. One of Z-1's steering gyroscopes has been plagued by technical problems but International Space Station Manager Tommy Holloway said the problems have been fixed and will not delay the launch date. ["Shuttle rolls out for Oct. mission," *Florida Today*, September 12, 2000, p 3A.]

SEPTEMBER 12: The STS-106 crew will have an extra day in space to move supplies into the International Space Station, NASA officials said. Atlantis is now due to return to Kennedy Space Center about 4:30 a.m. September 20 rather than 4:45 a.m. September 19. ["Atlantis crew gets extra day for mission," *Florida Today*, September 13, 2000, p 1A & 3A.]

◆   The crew of Discovery arrived at KSC for three days of pre-launch work. The STS-92 crew includes: Brian Duffy, commander; Pamela A. Melroy, pilot; mission specialists Leroy Chiao, Peter J. K. Wisoff, Michael E. Lopez-Alegria, and William S. McArthur. This flight will mark a historic milestone as it will be the 100th shuttle voyage into space since the fleet started flying in 1981. ["Discovery crew starts pre-launch activities," *Florida Today*, September 13, 2000, p 3A.]

SEPTEMBER 13: The Z1 truss was lifted into the Payload Changeout Room at Pad 39A at about 6 a.m. The first U. S. solar arrays will be temporarily attached to the Z1 in orbit to provide power for the initial work on assembly of the Integrated Truss Structure, the backbone of the International Space Station The arrays will later be permanently attached to the central truss segment SO, scheduled for launch in October 2001. ["Z1 truss at pad, set for Oct. 5 launch," *KSC Countdown*, September 14, 2000.]

◆   The crew of Discovery, STS-92, took turns driving an armored personnel carrier near Pad 39B as training for evacuation of the shuttle should an emergency occur while on the launch pad. If forced to evacuate the shuttle at the pad, the astronauts would have two choices after escaping down slidewires from the pad to the ground: to take shelter in a nearby bunker or to drive the personnel carrier (M113 tank) a mile to a helicopter pad where they would be airlifted to safety. In NASA's history, the personnel carrier has never been called into action for emergencies, according to NASA officials. ["Astronauts prepare for worst situations," *Florida Today*, September 17, 2000, p 8E & 4E.]

SEPTEMBER 15: Investigators have concluded that they don't know who put a bag of urine on the Titan launch pad Complex 40 at Cape Canaveral Air Force Station last month, but plans to improve the restroom facilities at the site are in the works. "The case is closed," said Ken Warren, the 45th Space Wing spokesman for the Air Force. ["Origin of urine bag on launch pad still mystery," *Florida Today*, September 16, 2000, p 11A.]

SEPTEMBER 16: The threat of Hurricane Gordon to the area had Kennedy Space Center preparing for the possible roll-back of STS-92 Discovery from Launch Pad 39A to the Vehicle Assembly Building depending upon Gordon's projected path. Launch team members loaded the Z-1 Truss into the Discovery's cargo bay, an operation that had been planned for September 19. The shuttle was covered with the rotating service structure. Other KSC workers spent the day boarding, shuttering and sandbagging. ["Gordon set to soak Brevard," *Florida Today*, September 17, 2000, p 1A & 2A.]

SEPTEMBER 17: NASA decided to keep shuttle Discovery (STS-92) on Launch Pad 39A to ride out the rain and wind. At 8:00 a.m., Kennedy Space Center operated under Hurricane Condition 3 due to Tropical Storm Gordon which means everything must be secured and shuttered. During this time KSC recorded wind guests of up to 50 MPH. Most workers were told not to report to work (Sunday). ["KSC was ready for Gordon," *Florida Today*, September 18, 2000, p 1A.]

SEPTEMBER 20: Space Shuttle Atlantis (STS-106) landed on Runway 15 at Kennedy Space Center's Shuttle Landing Facility at 3:56 a.m. despite early weather concerns. This marked the 23rd consecutive landing in Florida and the 30th landing of a shuttle at Kennedy Space Center. ["Focus turns to Discovery with end of Atlantis mission," *Florida Today*, September 21, 2000, p 1A & 5A. *Space Shuttle Mission Chronology 2000*, KSC Release No. 12-92, Revised November 2000, p 5.]

SEPTEMBER 22: A small leak in shuttle Discovery's propulsion system was discovered during a routine test of the shuttle's plumbing, which carries liquid hydrogen and oxygen to the three engines during launch. The leak is expected to be repaired quickly and should not delay the October 5 launch. Discovery is on the launch pad at Kennedy Space Center undergoing final preparation for its STS-92 mission to the International Space Station. ["Shuttle leak likely won't delay launch," *Florida Today*, September 26, 2000, p 1A.]

SEPTEMBER 28: NASA cleared Shuttle Discovery (STS-92) for launch on the 100th mission of the shuttle program. A leak that developed last week in the shuttle's propulsion system has been fixed, and shuttle managers declared the orbiter ready for liftoff. Launch is scheduled for 9:38 a.m. from Kennedy Space Center's Launch Pad 39A on October 5. ["Shuttle Discovery gets OK for Thursday liftoff," *Florida Today*, September 29, 2000, p 1A.]

DURING SEPTEMBER: Debate among safety officials at Cape Canaveral Air Force Station and Vandenberg AFB, CA, over an Air Force Space Command range safety shift that some managers believe could erode the "independent safety checks and balances" critical for safe U.S. human and robotic launches. The action involves moving elements of the Air Force range safety staffs at both the Cape and Vandenberg out of safety and into flight operations – the same organization charged with maintaining schedules and conducting launches. NASA has also been concerned about the Air Force shift as it relates to shuttle and NASA unmanned vehicle range safety oversight at both the Cape and Vandenberg. Although the changes put in place here in mid-August were in effect for the scheduled launch of the shuttle Atlantis late last week, the concern is more for the long term than any specific flight. "Safety cuts through every facet of launch operations, from the ground processing to the flight operations to public safety and down to bottom-line cost," said David Phillips, the NASA liaison between the Air Force operations community at the Cape and the NASA Kennedy Space Center. NASA has insisted on a new USAF/NASA Memorandum of Agreement (MOA) to specifically spell out critical responsibilities in light of the changes. ["Controversy erupts over USAF space range safety," *Aviation Week & Space Technology*, September 11, 2000, p 28-29.]

OCTOBER

OCTOBER 1: The seven members of STS-92 crew arrived at Kennedy Space Center Sunday, October 1. The historic 100th space shuttle flight is scheduled to launch at 9:39 p.m. Thursday. ["Shuttle crew arrives for historic mission," *Florida Today*, October 2, 2000, p 1A. "STS-92 crew arrives, facing 60% probability for launch Thursday due to weather," *KSC Countdown*, October 3, 2000.]

OCTOBER 2: The Joint Airlock Module, which will be attached to the International Space Station in May, successfully completed leak testing and was removed from a vacuum chamber in the Operations and Checkout Building at the Kennedy Space Center. The 6.5-ton airlock was installed in the payload transportation canister and moved to the Space Station Processing Facility where it will continue to undergo preflight processing for the STS-104 mission. ["Station's airlock aces test," *Florida Today*, October 3, 2000, p 1B. "ISS Hardware On the Move," *KSC Countdown*, October 3, 2000.]

OCTOBER 3: The countdown clocks at Kennedy Space Center were to have begun ticking early today toward NASA's 100th shuttle launch. However, there is no guarantee that the weather will cooperate. Hurricane Keith is expected to head north toward the Gulf of Mexico from its present position near the Yucatan Peninsula. There is a possibility the storm could send showers and clouds close to the launch site. ["100th launch, like first 99, is at mercy of the weather," *The Orlando Sentinel*, October 3, 2000, p A-4.]

◆ The launch countdown for STS-92 began at 12 a.m. As scheduled. [Bruce Buckingham. (2000). Kennedy Space Center Status Report [Online]. Available E-mail: domo@news.ksc.nasa.gov/subscribe shuttle-status [2000, October 3].]

◆ Pam Melroy, who will be piloting Discovery's STS-92 mission, is getting a lot of attention as only the third woman to pilot a shuttle. Melroy, an Air Force pilot, said she is more excited about making her first space flight than in becoming the third woman to pilot a shuttle. "I hope someday it won't be a big deal," Melroy said recently. ["Female shuttle pilot puzzled by attention," *Florida Today*, October 3, 2000, p 2A.]

◆ Russian space officials have recommended Russia dump the aging Mir space station into the ocean, a Cabinet official said Tuesday. Comments by Deputy Prime Minister Ilya Klebanov ran contrary to a statement issued Tuesday by Energia, the company which operates Mir. Klebanov said that the 14-year-old Mir no longer had any scientific value and was costing too much money. Energia, however, said Mir was ready to go on flying. Russia has been under pressure from U.S. space officials to let Mir fall back to Earth in the Pacific Ocean, and concentrate its funds on the International Space Station. ["Russia: It's time to dump aging Mir," *Florida Today*, October 4, 2000.]

OCTOBER 4: Two storm systems have NASA watching Florida and Texas as the countdown to Thursday's launch continues. There is a 40% chance that weather will postpone the mission to the International Space Station. The weather outlook is worse for Friday and Saturday. Officials are watching a tropical wave that formed over South Florida which is moving north. Hurricane Keith, which has weakened to a tropical depression, is forecast to make landfall 100 miles south of Texas, and should not affect Johnson Space Center's shuttle mission control. ["Weather could delay Discovery," *Florida Today*, October 4, 2000, p 1A. "Weather remains iffy for Discovery," *The Orlando Sentinel*, October 4, 2000, p A-6.]

◆ The Kennedy Space Center Visitor Complex announced a new tour option. The new 90-minute tour, "NASA Up Close," will include visits to the shuttle launch pads, the landing strip and press site. The new tour will cost $20 in addition to the $24 maximum access badge, and will take visitors along the same route used by astronauts on launch day. ["Visitor Complex enhances KSC tour," *Florida Today*, October 5, 2000, p 8A.]

OCTOBER 5: Two technical problems that developed late in Discovery's countdown will keep the shuttle grounded until at least Monday, October 9. During the standard pretanking meeting, shuttle managers decided to delay the launch of Shuttle Discovery by at least 24 hours. Engineers are concerned a bolt holding the orbiter to the external tank will not separate completely and endanger the shuttle. The bolt problem occurred during the shuttle Atlantis' September 8 launch but was not discovered until October 4, when film of that launch was being routinely analyzed. Engineers noted a 2 ¼-inch protrusion of the aft attach bolt following tank separation. At separation, a frangible nut inside the orbiter releases the 14-inch bolt which is supposed to fully retract into the external tank's bolt housing. On STS-106, that bolt did not fully retract. Since there is no way for workers to access the bolt, engineers are working with computer and mathematical models to determine whether or not the bolt poses a problem. Engineers are also evaluating a crucial valve in the shuttle's main propulsion system after it was slow to close during routine test today. [Bruce Buckingham. (2000). Kennedy Space Center Status Report [Online]. Available E-mail: domo@news.ksc.nasa.gov/subscribe shuttle-status [2000, October 5]. "Shuttle won't fly until Monday," *Florida Today*, October 6, 2000, p 1A. "Pair of technical glitches delays shuttle's launch until at least Monday," *The Orlando Sentinel*, October 6, 2000, p A-6.]

◆ Discovery's STS-92 mission to the International Space Station will install the 18,400 lb. Z-1 truss, which will be the basis for the station's main solar arrays, and which also contains gyroscopes and antennas. The Pressurized Mating Adapter (PMA-3) will also be added to the station, and the Common Berthing Mechanism (CBM) will be tested. There will be four, possibly five, extravehicular activities required to complete the installation of these critical station components. Discovery will be commanded by Brian Duffy (Col., USAF), and piloted by Pam Melroy (Lt. Col., USAF). Mission specialists include Leroy Chiao (Ph.D.), Bill McArthur (Col., USA), Jeff Wisoff (Ph.D.), Mike Lopez-Alegria (Cdr., USN) and Koichi Wakata, representing NASDA, the Japanese Space Agency. Discovery's five minute launch window, for the historic 100th space shuttle mission, will open at 9:39 p.m. EDT. During the 11-day mission, Wakata will use the shuttle's Canadian manipulator arm to attach the Z-1 truss and the PMA to the station's Unity module. The other mission specialists will be divided into two teams which will perform four planned space walks to make the connections for electricity and data communications between the two new components and the station. ["U.S./Japanese Crew Set to Install ISS Systems Truss," *Aviation Week & Space Technology*, October 2, 2000, p 73 & 74. "Discovery Liftoff Set for Oct. 5 to Begin Space Station Build-up," *NASA News Release #00-156*, September 29, 2000.]

◆   Florida Governor Jeb Bush and his wife Columba head the list of VIPs attending the launch of Discovery. Also included are singers James Brown and Patti LaBell; Jean Schultz, the widow of Peanuts creator Charles Schultz; actress June Lockhart; and Congressman Dave Weldon, who represents Brevard County. This is the first shuttle launch Bush has attended, and he is doing so at NASA's invitation. ["VIPs plan to witness historic launch," *Florida Today*, October 5, 2000, p 8A.]

◆   Why is STS-92 the shuttle's 100th mission? NASA scheduled the STS-92 mission about four years ago, and assigned it to take the Z-1 truss to the International Space Station. But two years of delays for the station meant the STS-92 mission would have to wait. Other missions with higher STS numbers, which didn't have anything to do with the station, were able to launch instead. NASA decided it would be too expensive and confusing to keep rearranging numbers every time a mission was delayed. ["Why is STS-92 the 100th mission?," *Florida Today*, October 5, 2000, p 8A.]

OCTOBER 6: Johnnie B. "J.B." Kump of Boeing Space Coast Operations at Kennedy Space Center will resign as director of communications and external relations, effective today. Kump, who has worked for Boeing for nearly three years, also served as district director for U.S. Rep. Dave Weldon, R-Palm Bay. Previous to that, Kump was chief of media relations for the former Lockheed Space Operations Co., now operating as United Space Alliance at Kennedy Space Center. ["Kump will leave Boeing to work for ministry," *Florida Today*, October 6, 2000, p 12C.]

OCTOBER 7: A faulty valve in shuttle Discovery's propulsion system was expected to be replaced by this morning. Meanwhile, engineers at Kennedy Space Center and Johnson Space Center are trying to determine whether the bolt problem which caused Thursday's scrub poses a threat to Discovery's mission. Weather will also play a role in when Discovery launches. High winds accompanying a cold front are forecast to move through the area today. Monday and Tuesday have only 30% chances of favorable weather. Wednesday is the last day Discovery can launch before the Eastern Range is turned over to the Air Force for an Atlas 2A launch planned for Thursday. ["Weather could delay Monday's shuttle launch," *Florida Today*, October 7, 2000, p 1A.]

◆   Far from the remote island of Kwajalein, in the Marshall Islands, Kennedy Space Center controllers, using computers at the Mission Director's Center at Cape Canaveral Air Force Station, are going to attempt to monitor and control all aspects of a launch. The launch, which had been set for today, was scrubbed, with no new launch date set. But when the launch does occur, it could herald a new path for KSC. A Pegasus rocket will be used, which is launched by being dropped from a converted airliner, fires its engine after release and can send small satellites into orbit. KSC controllers will monitor the rocket's health and tell the aircraft's pilots to drop the rocket for launch or to return with it to the base at the Kwajalein Missile Range in the South Pacific. Without the remote control capability, NASA would have to send 120 engineers and managers to Kwajalein. The rocket will send a NASA satellite, the High Energy Transient Explorer 2 into orbit. ["KSC to try remote controlled launch," *Florida Today*, October 7, 2000, p 5A.]

◆   Local and state activists will converge on Cape Canaveral Air Force Station this afternoon to protest the militarization of space. The protest comes the weekend before the launch of the 100th space shuttle mission and an Atlas rocket carrying an Air Force satellite. The Cape Canaveral protest is sponsored by the Florida Coalition for Peace and Justice, Pax Christi Florida and Global Network. ["Protestors object to military in space," *Florida Today*, October 7, 2000, p 5A.]

◆   Early Saturday morning workers at the Kennedy Space Center replaced a faulty valve which controls fuel flow in shuttle Discovery's propulsion system. The replacement valve passed leak checks, and workers are confident its replacement will operate properly. ["Discovery fuel valve replaced," *Florida Today*, October 8, 2000, p 1A.]

◆   One woman was arrested on trespassing charges during a peaceful protest against the militarization of space outside Cape Canaveral Air Force Station, Saturday. More than 60 people took part in the protest. ["Protest ends with 1 arrest," *Florida Today*, October 8, 2000, p 1B.]

OCTOBER 8: Shuttle managers decided Sunday afternoon that a bolt holding the shuttle to its external tank will not pose a threat to Discovery if the bolt does not fully retract soon after the shuttle reaches space and jettisons the tank. A bolt on a similar tank did not completely fall back into the tank during the last mission, but the shuttle Atlantis was not damaged. "The decision was made that it was safe to fly as is and that there was virtually no chance of the bolt causing…the external tank to hit the orbiter," NASA spokesman John Ira Petty said. ["Winds could further delay Discovery," *Florida Today*, October 9, 2000, p 1A. "Crosswinds could pose threat to Discovery's launch tonight," *The Orlando Sentinel*, October 9, 2000, p A-7.]

OCTOBER 9: At 9 a.m. this morning, mission managers postponed today's scheduled launch of Discovery's STS-92 mission to the International Space Station. A cold front passed through the area Saturday and brought winds which were as high as 44.6 knots at Launch Pad 39A, exceeding the limit of 42 knots. Due to the higher than allowable winds, the Gaseous Oxygen (GOX) vent arm or "beanie cap" could not be extended over the External Tank. The forecast calls for only a 30 percent chance that the weather will not interfere with a launch attempt tonight or Tuesday night. The forecast improves dramatically for Wednesday, with a 70 percent chance of favorable weather. Discovery has until Wednesday to launch before NASA has to give up the Eastern Range so the Air Force can launch and track an Atlas 2A rocket launch Thursday. ["Winds could further delay Discovery," *Florida Today*, October 9, 2000, p 1A. Bruce Buckingham. (2000). Kennedy Space Center Status Report [Online]. Available E-mail: domo@news.ksc.nasa.gov/subscribe shuttle-status [2000, October 9].]

◆   On Oct.9, an Orbital Sciences Corp. commercial Pegasus booster launched a U.S./French/Japanese gamma-ray burst detection satellite into space. The Pegasus with its High-Energy Transient Explorer (HETE-2) payload was dropped from OSC's L-1011 aircraft, which was staged the Kwajalein Missile Range in the South Pacific. Controllers at Kennedy Space Center managed countdown activities remotely from Florida. Data, voice and imagery from the L-1011 were routed to KSC by an unusual arrangement involving NASA, commercial and defense relay spacecraft. KSC Expendable Launch Vehicle Services personnel used consoles at the Cape Canaveral Air Force Station Hanger AE Mission Director's Center. This was NASA's first remotely managed launch. ["ELV update," *KSC*

Countdown, October 12, 2000. "Kennedy manages Pacific launch," *Aviation Week & Space Technology*, October 16, 2000, p 61.]

OCTOBER 10: Tonight's launch attempt for STS-92 was postponed during the T-20 built-in hold. Liftoff from the Kennedy Space Center was scheduled for 7:40 p.m., however an 8-ounce metal pit pin discovered late in Tuesday's countdown caused another launch delay for Discovery's STS-92 mission. High wind conditions and low clouds, which had caused the launch scrub Monday night, had improved enough to try to launch Tuesday. Tuesday's 24-hour delay was due to a member of the ice and debris inspection team seeing an 8-ounce metal locking pin on a strut between the orbiter and the external tank. An inspector using binoculars discovered the 4-inch long pin, attached to a 12-inch wire cord. The pin is believed to have fallen from a work platform or handrail. Launch officials feared the pin could fall off the strut during launch and damage the shuttle or tank or get sucked into one of the main engines at ignition. Shuttle officials called for the scrub after concluding the pin could not be removed quickly or safely knocked from its perch. Shuttle managers will investigate what happened to allow the pin to fall and institute additional measures to make sure it does not happen again. The weather forecast for Wednesday is for a 60 percent chance of favorable weather, and Thursday the chance of favorable conditions increases to 70 percent. Launching on Thursday became a possibility when an Atlas 2A launch scheduled for Thursday from Cape Canaveral was delayed. ["Windy day spurs new delay for Discovery," *The Orlando Sentinel*, October 10, 2000, p A-3. "Discovery to make 3rd launch attempt," *Florida Today*, October 10, 2000, p 1A. Bruce Buckingham. (2000). Kennedy Space Center Status Report [Online]. Available E-mail: domo@news.ksc.nasa.gov/subscribe shuttle-status [2000, October 10]. "8-ounce pin delays Discovery," *Florida Today*, October 11, 2000, p 1A & 3A. "Pin drop delays shuttle," *The Orlando Sentinel*, October 11, 2000, p A-7.]

◆ The launch of an Atlas 2A rocket from Cape Canaveral Air Force Station, scheduled for Thursday has been postponed indefinitely because of an electrical problem with its payload. A problem was discovered with a timer during a routine test of the DOD communications satellite. It will take about a week for a team to investigate the problem and determine the next step. ["Electrical glitch postpones Atlas," *Florida Today*, October 11, 2000, p 3A.]

OCTOBER 11: The Rotating Service Structure was moved back around Discovery at midnight and NASA engineer Jorge Rivera, who originally spotted the pin which caused Tuesday's launch scrub, retrieved it. The pin was embedded about one-eight of an inch into the tank's orange foam insulation. Discovery roared into space from Pad 39A, at 7:17 p.m., this evening on STS-92's history making 11-day mission to the International Space Station. The 100th shuttle launch from the Kennedy Space Center came after three launch delays for an assortment of technical, safety and weather reasons. The Discovery will catch up to the station 235 miles above the Earth, and dock with it about 1:45 p.m. Friday. The shuttle's crew will deliver and attach the Z-1 truss and an 8-foot-long docking tunnel to the station. Four spacewalks will be required to complete the two construction tasks on the rapidly growing space station. After the construction phase of the mission is completed and maintenance and supply tasks are finished, the crew will undock at about 9:30 a.m. October 20, and land back at KSC on October 22, at 2:10 p.m. This launch was only the second time in 12 years that a 24-hr. launch turnaround was achieved with the complex service structure rollback and retraction. ["100th soars toward station," *Florida Today*, October 12, 2000, p 1A & 4A. "This time is the charm for shuttle Discovery," *The Orlando Sentinel*, October 12, 2000, p A-1 & A-6. "Discovery ISS assembly underway," *Aviation Week & Space Technology*, October 16, 2000, p 61.]

◆   A NASA engineer who spied a 4-inch pit pin wedged against the shuttle Discovery's fuel tank will get a medal for quite possibly saving the lives of the seven astronauts of the STS-92 mission. Jorge Rivera, 43, said he was just doing his job. NASA halted Tuesday's countdown after Rivera, using binoculars, spotted the pin. Rivera was also responsible for Discovery's first launch delay. He was examining launch films of the STS-106 mission in September, when he noticed that one of the bolts on the external fuel tank was sticking out two inches. It should have been fully retracted. NASA called off the Oct. 5 launch attempt until engineers determined the bolt malfunction could occur on Discovery and whether it could cause the fuel tank to slam into the shuttle. It was decided that was unlikely. On Wednesday morning, Rivera retrieved the pin, which was embedded about one-eight of an inch into the tank's orange foam insulation. Because of Rivera's back-to-back finds, NASA's Administrator, Daniel Goldin said "I can't manufacture medals overnight, but we're going to give you the NASA exceptional achievement medal as a team to show you did the right thing." Rivera and the rest of the debris-inspection team are normally looking for ice on the external tank, which is filled with super-cold fuel. Launch director Mike Leinbach gave his Launch Director's Plaque award for this mission to Rivera, saying "Jorge is one of KSC's finest, and I'm extremely proud of him." ["100th soars toward station," *Florida Today*, October 12, 2000, p 1A & 4A. Web posted. (2000). NASA lauds hero engineer [Online]. Available WWW: http://www.usatoday.com/news/ndsthu03.htm [2000, October 12]. "NASA thanks engineer for discoveries," *Florida Today*, October 13, 2000, p 1B.]

◆   Weather plays an important role in whether or not a shuttle can launch from or land at Kennedy Space Center. It cannot be raining at the launch pad or within the shuttle's flight path, winds cannot exceed 34.5 mph for most missions and there can be no lightning within 10 nautical miles of the launch pad or within the flight path within 30 minutes of launch. Rain is the shuttle's worst enemy. At the high speed the shuttle attains after launch, raindrops can damage the orbiter's sensitive tiles which protect it from the extreme heat of re-entry. Winds are a big factor at the pad and the 3-mile-long Shuttle Landing Facility. Because the shuttle is essentially a glider when it lands, it has only one chance to land properly. Winds and clouds can interfere with an emergency landing back at KSC should the shuttle encounter a problem within the first 4 minutes and 20 seconds of flight. Weather conditions at emergency landing sites have to be good as well. At least one of the transatlantic abort sites must have good weather in case the shuttle must land before reaching orbit, and weather at Edwards Air Force Base in Calif., and at White Sands, N.M., must also be acceptable in case the shuttle has to land after just one orbit. Temperatures must also be within certain limits for fueling to be carried out. ["Weather just fine for launch," *Florida Today*, October 12, 2000, p 4A.]

OCTOBER 12: The Senate voted by a margin of 87-8 to give NASA its biggest budget boost in years, a $683 million increase from last year's level. This would mark the end of a seven-year slide in NASA's finances. NASA's appropriation is contained in the annual bill providing funds for the departments of Veterans Affairs, and Housing and Urban Development and about a dozen other independent agencies. At NASA's request lawmakers shifted $37 million from the human space-flight account to provide more money for the Mars 2003 lander. An additional $40 million is set aside so NASA can begin developing an alternate escape module for the International Space Station, as a hedge against expected further difficulties with Russian hardware and services. The space station is financed at $2.114 billion, $211 million lower than last year, and the Mission to Mars is fully financed at the agency's requested level of $420 million. The House is expected to take up the measure next week. ["Senators OK book in budget for NASA," *The Orlando Sentinel*, October 13, 2000, p A-5.]

◆ Launch pad inspectors found no abnormal damage following Discovery's launch last night. The solid rocket recovery ships report both boosters are in good condition. [Bruce Buckingham. (2000). Kennedy Space Center Status Report [Online]. Available E-mail: domo@news.ksc.nasa.gov/subscribe shuttle-status [2000, October 12].]

OCTOBER 13: Shuttle Discovery's astronauts overcame navigation equipment trouble and successfully docked with the International Space Station at 1:45 p.m. An antenna used for radar and sending video images to Earth malfunctioned Thursday. The problem was linked to an oscillator which slowed transmission of electronic files. This was the first time in almost 20 years a shuttle docked with another spacecraft without the use of the main antenna for radar. The crew used laser devices to measure Discovery's distance from the station and to determine how fast the shuttle was approaching. The crew entered the station's Unity module to prepare for Sunday's spacewalk. ["Discovery docks with space station," *Florida Today*, October 14, 2000, p 1A. "Discovery rendezvous successful," *The Orlando Sentinel*, October 14, 2000, p A-3.]

OCTOBER 14: Discovery's crew overcame several technical problems which made the addition of a new segment to the International Space Station more difficult. Japanese astronaut Koichi Wakata used the shuttle's robot arm to inch the 18,400-pound Z-1 truss out of the shuttle's payload bay and into place on the station. Work was delayed about 3 ½ hours by a short circuit in the shuttle. An electrical spike blew a circuit breaker, which in turn shut off a cargo bay camera and a laptop computer video system used to track the installation. It also killed one of two cameras on the end of the robot arm. The computer system, which tracks an array of dots on the station's modules, was important because Wakata did not have a direct view of the Z-1 moving toward the station. The crew and controllers at Johnson Space Center had simulated such failures in training, and the crew quickly replaced several components in the computer video system and hooked it up to a new power source. While Wakata had the task of using the robot arm, Mike Lopez-Alegria looked through a porthole inside the Unity module, looking out at the Z-1 as it approached. He used a flashlight at one point to make sure the module was lined up. Pilot Pam Melroy used a laptop computer to activate latches on the station that reached out and grabbed the Z-1 and bolted it into place. The $273 million box-shaped truss will become the hub of the station's girder-like framework. It contains the station's main communications antenna as well as four stabilizing gyroscopes needed to help the station conserve propellant. ["Station gets new segment," *Florida Today*, October 15, 2000, p 6A. "Astronauts have one of those days," *The Orlando Sentinel*, October 15, 2000, p A-3.]

OCTOBER 17: Astronauts Bill McArthur and Leroy Chiao worked outside shuttle Discovery for six hours and 48 minutes on their second spacewalk of the STS-92 mission to the International Space Station. Two 129-pound power converters were released from the shuttle's payload bay, and installed on the Z-1 truss. The power converters are necessary to convert the electricity produced by the station's solar arrays into a useable voltage. The solar arrays will be added to the station during the upcoming STS-97 mission. Just moments after Bill McArthur emerged from the shuttle, a cap for a depressurization valve floated away. The aluminum cover, about the size of a gas cap on a car, bounced against the space station, then against the shuttle's robot arm. Then it was gone, an addition to the collection of junk which orbits the Earth. Mission Control said the lose of the cap was no reason for concern. ["Astronauts attach power converters," _Florida Today_, October 18, 2000, p 1B. "Cap is lost in orbit as devices go on station," _The Orlando Sentinel_, October 18, 2000, p A-5.]

OCTOBER 18: Jet backpacks, which will be used on future missions if an astronaut ever becomes incapacitated or floats away from the shuttle or the International Space Station, were tested by astronauts Jeff Wisoff and Michael Lopez-Alegria. With their station construction tasks completed, they took turns testing the new jetpack in and near Discovery's payload bay. Because the construction part of their spacewalk had lasted longer than expected due to a minor tool problem, the astronauts were unable to practice how one astronaut using a jetpack would carry an incapacitated spacewalker back to the shuttle In the 1980s, astronauts used a larger Manned Maneuvering Unit to move around untethered to a space shuttle. But these units were mothballed because they were too expensive and heavy. Previous rescue plans to recover an astronaut who had floated away from the shuttle, had the shuttle's crew fly the shuttle to where the astronaut was floating, and pick him up. Now with visits to the station, shuttles are docked, and it would be impossible to retrieve an astronaut who is floating free. With one-hundred sixty spacewalks required to build the space station, the new jetpacks are the solution to this problem. ["Spacewalkers test jetpack system," _Florida Today_, October 19, 2000, p 3A. "Astronauts find these backpacks can really fly," _The Orlando Sentinel_, October 19, 2000, p A-4.]

◆ NASA Administrator Daniel S. Goldin today recognized members of the Space Shuttle ice and debris inspection team at Kennedy Space Center, FL, for their keen safety observations prior to the recent launch of Space Shuttle Discovery. Gregory N. Katnik and Jorge E. Rivera of NASA received the Agency's Exceptional Achievement Medal, while Michael Barber, John B. Blue and Thomas F. Ford of United Space Alliance and D. Scott Otto of Lockheed Martin Space Services Company received the NASA Public Service Medal. On October 10, the team found a stray 4-inch pin near the shuttle's external fuel tank while using binoculars to scan launch pad 39A several hours before launch of STS-92. The discovery delayed the shuttle mission 24 hours, allowing the team to retrieve the pin and clear the shuttle for a safe launch. If not removed, the pin could have damaged the space shuttle's thermal protection system or could have been sucked into one of the main engines during launch, causing damage there. "Safety is NASA" number one priority, and this team exemplifies our commitment. The Agency is extremely proud of the inspection team for placing astronaut safety above adherence to launch schedules," said Goldin. ["Space Shuttle Inspection Team Rewarded for Its "Eagle Eyes," _NASA News Release #00-165_, October 18, 2000.]

◆ Boeing showed off a crucial piece of hardware for its fledgling Delta 4 rocket program: a high-tech, ocean-going hangar that doubles as a ship. The Delta Mariner is designed to carry three Delta 4 rocket cores or first stages, plus a payload fairing and a pair of upper stages in a half-million cubic feet of cargo space. As big as a football field, the Delta Mariner was built to transport rocket stages from Boeing's Decatur, Ala., facility to Port Canaveral, and eventually to Vandenburg Air Force Base, Calif. When the vessel reaches the Port, the stages will be offloaded on a specially built 80-ton transporter, then driven overland to the processing facility at Cape Canaveral Air Force Station. The first launch of the Delta 4 from CCAFS is scheduled for late 2001. Boeing will use Delta 4 rockets to support both military and commercial customers. ["Delta ship to help carry rocket parts," *Florida Today*, October 19, 2000, p 3A.]

◆   Kennedy Space Center and the 45th Space Wing recently celebrated the 3rd annual *Super Safety and Health Day*. All normal work activities, with the exception of mandatory services such as fire and security, were suspended, permitting thousands of NASA and Air Force employees and contractors to participate in a full day of informative activities. The theme, "Safety and Health...A Working Relationship," which focused on the crucial role safety and health plays in the overall effectiveness and success of mission goals and objectives. The afternoon program included a keynote address by Dr. Beck Weathers, a survivor of the 1996 climbing tragedy on Mt. Everest. ["KSC and 45th Space Wing observe safety and health day," *The Brevard Technical Journal*, November, 2000, p 6.]

OCTOBER 19: Discovery astronaut Jeff Wisoff went from spacewalker to plumber overnight. Pilot Pam Melroy, was performing the daily routine of compacting the waste in the shuttle's toilet, when it malfunctioned. With other crew members busy opening hatches on the International Space Station, the task of scooping out the solid waste fell on Wisoff. There was not any further word about the problem until Melroy's voice came over the communications line. "Well, Jeff is more of a hero than most people will appreciate. We got it taken care of and everything is back to normal," Melroy said. Toilets have been a problem during previous shuttle missions and on Mir. A new state-of-the-art toilet is aboard the ISS, but it is not in a module the crew was scheduled to be in nor is it fully assembled. [Web posted. (2000). "Toilet Malfunction [Online]. Available WWW: http://abcnews.go.com/ [2000, October 19].]

◆   The Air Force launched a defense communications satellite from Cape Canaveral Air Force Station tonight at 8:40 p.m., in another step toward upgrading an aging fleet. The Defense Satellite Communications Systems spacecraft was carried into orbit on an Atlas 2A rocket from launch pad 36A. The launch was pushed back from 7:36 p.m. because of a series of problems including a barge in the launch hazard area, a computer file set for a larger payload that what the rocket carried and fear that a fuel line would not retract before launch. The $200 million, 2,700-pound satellite, built by Lockheed Martin Space Systems, is to be part of a ten-satellite constellation for secure voice and data transmission for the military. This satellite will replace one which was launched from the shuttle Atlantis' payload bay 15 years ago. Launch was delayed for a week because of a problem with the spacecraft's initiation timer. ["Atlas launch is set for tonight," *Florida Today*, October 19, 2000, p 3A. "Atlas 2A launches despite problems," *Florida Today*, October 20, 2000, p 3A.]

◆   The House approved NASA's first spending increase in seven years, giving the space agency a $14.285 billion budget for the fiscal year that began Oct. 1. The spending bill fully funds shuttle operations, salaries and expenses at the levels requested by NASA. "We received what the president requested," said Sarah Keegan a NASA spokeswoman. "That's a good thing from our perspective." The money represents an increase of $683 million – 4.4 percent – over last year's level, but about $250 million of it was due to special projects requested by members of Congress. At NASA's request the spending bill moves approximately $75 million from various accounts to supplement the Mars 2003 Lander program. Last year's loss of two Mars probes has prompted NASA to spend more to ensure future missions arrive safely. The legislation also restores $290 million for continued development of a second-generation reusable launch vehicle which would eventually replace the shuttle. Funding for the International Space Station is $2.1 billion, the same as last year. Another $90 million is approved for a year's work on the "lifeboat" NASA wants to send to the ISS. The vehicle would supplement the Russian Soyuz vehicles that will serve as the primary escape craft in case of emergency. Lawmakers also agreed to spend $3 million to complete the design of a "bioastronautics" facility at the Johnson Space Center to study the effects of long-duration space flight. Kennedy Space Center's share of the pie is $1.6 billion. Lisa Malone, KSC spokeswoman said "We expect we would get equal to that and would hope for little more. That will be adequate to do the job here at KSC." The budget also fully pays for a shuttle upgrade program that will improve safety and performance, said Rep. Dave Weldon, R-Palm Bay. ["House OKs NASA budget hike," *Florida Today*, October 20, 2000, p 1A. "NASA budget gets a bit of a boost," *The Orlando Sentinel*, October 20, 2000, p A-19.]

◆ The University of Central Florida, leading an alliance of universities and colleges, expects to receive a boost in money and influence from state officials toward creating a major research center at Kennedy Space Center and Cape Canaveral Air Force Station. The center would build on programs UCF is leading there, looking into the science and engineering of spaceflight. "My target is to have 600 resident students on this campus," said Ronald L. Phillips, director of UCF's Florida Space Institute. "We expect this to be a joint campus for the entire university system, a residential campus where university students come here and do their academic work and do research. We're looking for 30 resident faculty to be here," Phillips said. On Thursday the Florida Council of University Presidents supported a plan to make the UCF Florida Space Institute and the University of Florida co-leaders of a statewide alliance for space research and education. The UCF Institute already leads an alliance of 10 schools, involving about 400 students and 10 full-time faculty. NASA and the Air Force have provided them laboratory, office and classroom space, as well as widespread access to launch facilities and government labs. Edward Ellegood, director of policy and program development at Spaceport Florida Authority, Florida's space-development agency, said "It's a long time coming. Kennedy Space Center is committed to redefining itself, expanding its role as a research-and-development center, beyond just a launch-operations center." Florida is also preparing to build a new research building, jointly used by NASA, industry and UF at the Cape, and operating under the alliance. ["UCF aims for the stars with center," *The Orlando Sentinel*, October 20, 2000, p D-1 & D-7.]

OCTOBER 20: Because it took longer than expected for Discovery's crew to move computers and camera gear from the shuttle into the International Space Station yesterday, the hatches to the station will not be closed until this morning. Before they left the crew tested newly installed equipment and checked for mold. The undocking at 11:08 a.m. marked the end of an ambitious and successful visit to the station, the last before the first permanent crew moves into the 13-story-tall facility in early Nov. Pilot Pam Melroy steered Discovery away from the station this morning and the crew is preparing for their Sunday return to the Kennedy Space Center. The landing is scheduled for 2:14 p.m., Sunday, although NASA is watching for possible high winds. ["Crew to close space station hatches today," *Florida Today*, October 20, 2000, p 3A. Bruce Buckingham. (2000). Kennedy Space Center Status Report [Online]. Available E-mail: domo@news.ksc.nasa.gov/subscribe shuttle-status [2000, October 20].]

◆ NASA Administrator Daniel S. Goldin praised the newly passed appropriations bill, which authorized $14.285 billion for the Agency. "Thanks to the efforts of key members of the House and Senate, and with the support of the Administration, this measure provides an excellent budget for NASA," Goldin said. ["NASA Administrator Reacts to FY 2001 Appropriation," *NASA News Release #00-159*, October 20, 2000.]

OCTOBER 22: The space shuttle Discovery was scheduled to end its historic STS-92 mission to the International Space Station at Kennedy Space Center this afternoon at 2:13 p.m. However, NASA managers concerned that cross winds higher than the 15 knot limit at the runway postponed the landing. The high winds are not expected to diminish Monday or Tuesday. Rain expected at Edwards AFB today and Monday, will make the landing strip there unavailable until Tuesday. The crew has enough supplies on board to stay on orbit until Wednesday. NASA wants Discovery to land at KSC to avoid the $1 million and week of work it takes to bring the shuttle back to KSC from Edwards. ["Winds may delay shuttle landing," *Florida Today*, October 22, 2000, p 1A. Buckingham. (2000). Kennedy Space Center Status Report [Online]. Available E-mail: domo@news.ksc.nasa.gov/subscribe shuttle-status [2000, October 22].]

◆ The U.S. Space Walk of Fame Foundation held a reunion and picnic for retired space workers at Fox Lake Park, Sunday. The foundation, which was created in 1992, focuses attention on the people who were less visible than the astronauts and higher profile personalities, but who played a vital role in the beginning of this country's space programs. ["Space workers reunite," *Florida Today*, October 23, 2000, p 1B.]

◆ Orbiters Endeavour and Atlantis are being prepared for upcoming launches in November and January 2001. In Orbiter Processing Facility (OPF) bay 2, preparations are under way for Endeavour's rollover to the Vehicle Assembly Building. In OPF bay 3, Atlantis's window No. 8 has been replaced and testing of the main propulsion system is in work. Workers are also preparing to remove the right hand orbital maneuvering system pod. [*KSC Countdown*, October 22, 2000.]

OCTOBER 23: NASA flight controllers in Houston, TX, waived off all shuttle landing opportunities as the weather at Kennedy Space Center and at Edwards Air Force Base in Calif., were unfavorable. [Bruce Buckingham. (2000). *Kennedy Space Center Status Report* [Online]. Available E-mail: domo@news.ksc.nasa.gov/subscribe shuttle-status [2000, October 23]. "Discovery makes 2nd landing try today," *Florida Today*, October 23, 2000, p 1A. "Shuttle landing delayed," *The Orlando Sentinel*, October 23, 2000 p A-1.]

OCTOBER 24: After Monday's high winds at Kennedy Space Center and high winds and showers at Edwards Air Force Base in California forced shuttle Discovery to remain in space another day, Discovery completed its STS-92 mission, landing at Edwards, AFB, at 5 p.m. EDT. It will now take about a week, and $1 million to return Discovery to KSC. This was the first shuttle landing in California since March 1996. Of the programs 98 landings to date, 45 have ended in California and 52 have landed in Florida. One shuttle landed at White Sands, N.M. in 1982. Since March 1996, 23 consecutive flights have returned to KSC. ["Wind, rain keep shuttle in orbit," *Florida Today*, October 24, 2000, p 1A. "California shuttle landing looks likelier," *The Orlando Sentinel*, October 24, 2000, p A-3. "Discovery lands in California," *Florida Today*, October 25, 2000, p 1A & 2A. Web posted. (2000). Shuttle crew readjusts to Gravity While Shuttle Awaits Transport [Online]. Available WWW: http://abcnews.go.com/sections/science/ [2000, October 25]. Bruce Buckingham. (2000). Kennedy Space Center Status Report [Online]. Available E-mail: domo@news.ksc.nasa.gov/subscribe shuttle-status [2000, October 24].]

◆   While waiting for Discovery to return to Kennedy Space Center, workers there were busy with shuttle Endeavour. The shuttle was rolled out of the Orbiter Processing Facility (OPF) bay 2, and into the Vehicle Assembly Building transfer aisle at 4:00 a.m. this morning. Workers in the VAB are preparing to lift Endeavour into high bay 1 where it will be mated to the external tank and solid rocket boosters. ["Discovery lands in California," *Florida Today*, October 25, 2000, p 2A. Bruce Buckingham. (2000). Kennedy Space Center Status Report [Online]. Available E-mail: domo@news.ksc.nasa.gov/subscribe shuttle-status [2000, October 25].]

◆   Some of the crew for mission STS-107 were at SPACEHAB recently to check out some of the equipment on their research mission, scheduled for July 2001. The visiting astronauts included Mission Specialists Michael Anderson, Ilan Ramon and Kalpana Chawla. This week members of the STS-98 crew will be at KSC for Crew Equipment Interface Test activities. STS-98, scheduled to launch on Jan. 18, will be the seventh flight to the International Space Station. ["Other mission activity," *KSC Countdown*, October 24, 2000.]

◆   The Baikonur Cosmodrome in the former Soviet republic of Kazakstan suspended work Tuesday, observing a day of mourning for the victims of two Soviet-era launch pad explosions that killed about 100 people. Both accidents involved intercontinental ballistic missiles which exploded on their launch pads during the early 1960s. In Moscow, top space officials honored victims of the disasters by laying flowers at the Kremlin wall. The annual commemoration interrupted preparations for the planned October 31 launch of the first permanent crew to the International Space Station. ["Russian space center mourns workers killed in two space disasters," *Florida Today*, October 25, 2000, p 2A.]

OCTOBER 25: Hoping to get tourists to spend more time in Brevard County, the Kennedy Space Center Visitor Complex plans to start selling travel packages this fall. The packages would include tickets to the Space Center with local hotel stays and tickets to other area attractions. KSC draws more than 2.8 million visitors a year, and is the second largest tourist attraction behind area beaches, but an estimated 70% of KSC's visitors leave Brevard without staying overnight. Now Delaware North Park Services of Spaceport, Inc., plans to begin offering the travel packages in November or December. Tourism is already more than a $600 million-a-year industry, but a lack of local travel packages has been a weakness. When Melbourne airport officials recently traveled to Europe to stir up business, they were told by tour operators they would consider flying tourists into Melbourne if local travel packages were available. Most of the KSC's visitors come from outside Florida and outside the U.S., and they come to the Space Center for a day. Tourism officials want to change that so tourists will spend time at Brevard's beaches and other attractions. ["KSC plan may boost tourism," *Florida Today*, October 25, 2000, p 1A & 3A.]

◆ Delaware North Parks Services of Spaceport is more than five years and $130 million into a tourism expansion plan for Kennedy Space Center, and the company says more is to come. Delaware North plans to add attractions at KSC, some scheduled to open within months, the others taking years to develop. Current projects include: Development of "Mad Science," a live show about the challenges of long duration space travel, featuring 3-D effects and pyrotechnics. It should open in December or January. Preparing to build a new access road to the Visitor Complex and a nearby proposed research park. The project to start next year, will offer 24-hour access from State Road 3 and State Road 405. Discussions to have film maker James Cameron – director of "Titanic" and the "Terminator" movies produce a space-related film for the Visitor Complex. Delaware North plans to spend about $100 million on Space Center attractions during the next five years. ["Additional attractions planned for KSC," *Florida Today*, October 25, 2000, p 3A.]

OCTOBER 26: A detailed review of a Space Shuttle Main Engine test mishap, June 16, at NASA's Stennis Space Center, MS, has revealed that special tape was left behind inside the engine during processing, contaminating the system. Robert Sackheim, Assistant Director for Space Propulsion at NASA's Marshall Space Flight Center, Huntsville, AL, to assess the main engine test mishap. The investigation team found that nearly 24 square inches of tape, routinely used as a temporary closure or protective barrier during main engine processing and assembly, had been inadvertently dropped into the fuel system. Despite normal inspections, it went unnoticed before the engine's test firing. The tape fell on the fuel and oxygen preburner injectors, with the majority of the tape in the fuel preburner. The tape blocked the multiple fuel-inlet holes causing an oxygen-rich mix, which rapidly increased temperature beyond the engine's operating limits, and melted some components upstream of the engine's fuel pump. The engine controller performed as designed, shutting down the engine when it sensed a temperature that exceeded the safe limits sat by engineers for this test. The engine was a development unit rather than a flight configuration. Sackheim's team found inadequate procedures for use of loose materials, and the use of the tape as a barrier material provides the opportunity for material to be left in an engine. Recommendations for improvement of procedures to prevent another occurrence of this problem were included in the team's report. ["Shuttle Main Engine Test Investigation Points to Fuel System Contamination," *NASA News Release #00-170*, October 26, 2000.]

◆ U.S. Astronaut Bill Shepherd and Russian cosmonauts Sergei Krikalev and Yuri Gidzenko left Moscow for the Baikonur cosmodrome in the former Soviet republic of Kazakstan for a planned launch to the International Space Station, Tuesday (October 31) at 2:53 a.m. ["Russia funds Mir, mum on its fate," *Florida Today*, October 27, 2000, p 6A.]

OCTOBER 27: NASA officials met with representatives of 11 different aerospace companies to explain the licensing process for commercial use of the Personal Cabin Pressure Altitude Monitor, designed at Kennedy Space Center, by Jan Zysko. It is meant to warn aviators when an airplane cabin pressure approaches a hazardous level. Hypoxia, caused by flying above 10,000 feet, is when there is a lack of oxygen in the blood and tissues. Zysko hopes his invention will prevent future accidents caused by hypoxia, such as the one in which golfer Payne Stewart and several other people died. ["NASA monitor may save fliers," *Florida Today*, October 29, 2000, p 1E.]

OCTOBER 29: Flight controllers in Houston and Moscow spent Sunday practicing for the arrival of the first crew to the International Space Station. Meanwhile the Russian Soyuz rocket that will carry the crew rode a train to its launch pad in Kazakstan. The rocket was raised into place, and a cocoon of metal gantries was raised around it for the final preparations before launch. Soyuz capsules are also used as lifeboats for the station in case of a medical emergency or an evacuation. ["Crew gets in shape for voyage to station," *Florida Today*, October 30, 2000, p 1A.]

OCTOBER 30: NASA and Lockheed Martin Space Systems have completed negotiations on a six-year contract for production of 35 super lightweight space shuttle external propellant tanks. Valued at $1.15 billion, agreement includes production and test of the 154-ft.-long tanks as well as operation of NASA's Michoud Assembly Facility in New Orleans. The tanks are about 7,500 lb. lighter in weight than the tanks they replace as a result of changes in design and material. The super lightweight tanks, first flown in 1998, are made of aluminum lithium. This sixth production of tanks will be the first comprised totally of Super Lightweight Tanks. The first tank of the sixth production is scheduled for delivery to Kennedy Space Center in 2002. ["Lightweight tanks," *Aviation Week & Space Technology*, November 6, 2000, p 23. "NASA Awards $1.15 Billion Contract for Shuttle External Tanks," *NASA News Release #C00-p*, October 30, 2000.]

OCTOBER 31: Tucked into the 25-foot-long Russian Soyuz spacecraft, Expedition 1 Commander U.S. Astronaut Bill Shepherd, and Russian cosmonauts Yuri Gidzenko and Sergei Krikalev lifted off from the Baikonur Cosmodrome in Kazakstan, heading for the International Space Station. After liftoff, they deployed antennas and other mechanisms that will allow the spacecraft to automatically join itself to the station. If the automatic docking fails, Gidzenko will take control of the craft and docking. ["Station crew prepares to dock," *Florida Today*, November 1, 2000, p 1A.]

◆   Bad weather in California and a problem with a bolt pushed back the first leg of shuttle Discovery's return to Florida. Discovery, which had just ended a 13-day assembly mission to the International Space Station, was to begin its trip back to Kennedy Space Center on the back of a specially equipped Boeing 747. High winds prevented workers from installing a protective tail cone over the shuttle's back compartment. Then there was a problem with one of the eight bolts which connect the cone to Discovery. Once it arrives back at KSC Discovery will be prepared for its next launch, on Feb. 15, which will take it back to the station. ["Bad weather, bolt problem keep shuttle in California," *Florida Today*, November 1, 2000, p 5A. Bruce Buckingham. (2000). Kennedy Space Center Status Report [Online]. Available E-mail: domo@news.ksc.nasa.gov/subscribe shuttle-status [2000, October 31].]

◆   Space shuttle Endeavour began its journey to Launch Pad 39B about 7 a.m. today. The assembly flight to the International Space Station is set for 10:05 p.m. Nov. 30. As the crawler transporter started up the pad slope about 12:30 p.m., workers noticed a crack in a link or cleat, in its tread. Each of the 57 links in a tread weighs about one ton. NASA managers stopped the vehicle on the ramp, and moved it back to level ground. The tread was repaired and the shuttle was moved the rest of the way without incident. Endeavour was hard down at the pad at 5:30 p.m. During Endeavour's STS-97 mission, the orbiter will deliver the massive solar arrays which will supply electricity to the station. [" Endeavour arrives at pad," *Florida Today*, November 1, 2000, p 5A. Bruce Buckingham. (2000). Kennedy Space Center Status Report [Online]. Available E-mail: domo@news.ksc.nasa.gov/subscribe shuttle-status [2000, October 31]. Bruce Buckingham. (2000). Kennedy Space Center Status Report [Online]. Available E-mail: domo@news.ksc.nasa.gov/subscribe shuttle-status [2000, November 1].]

DURING OCTOBER: The 100th plaque for the 100th shuttle mission will be added to walls in the Launch Control Center at Kennedy Space Center, and the Mission Control Center at Johnson Space Center when Discovery lifts off on the STS-92 mission. It will join the 31 plaques from the Mercury, Gemini, Apollo and Skylab programs, and the previous 99 shuttle flights. ["Shuttle widens window to space," *Florida Today*, October 1, 2000, p 1A & 7A.]

◆　The Medevac Oxygen System, developed at Kennedy Space Center, is to be used by the U.S. Air Force through a technology transfer agreement with the KSC Technology Programs and Commercialization Office. The system, created by NASA biomedical electronics technician Barry Slack, was originally designed to provide therapeutic oxygen supply to astronauts being flown aboard the C-130 aircraft in case of a forced landing at a Space Shuttle Transatlantic Abort Landing (TAL) site. The system is now being tested for planned incorporation into the U.S. Air Force Air Mobility Command for use in C-130s and C-141s. The U.S. Navy and U.S. Coast Guard are also considering using the Medevac Oxygen System aboard their aeromedical evacuation aircraft. Slack who works in the Biomedical Lab at KSC, equips Black Hawk helicopters with medical supplies to support potential medical needs for astronauts during launch and landing. ["KSC transfers oxygen delivery system technology to USAF," *The Brevard Technical Journal*, November, 2000, p 6. "Technology transfer," *KSC Countdown*, October 17, 2000.]

◆　Beal Aerospace Technologies, Inc., last week ceased operations in the wake of NASA plans to subsidize development of human-rated vehicles the company's chief official claims would compete directly with its BA-2C heavy-lift booster system. Andrew Beal, chairman, said development of a reliable, low-cost launch system such as the BA-C2 is "Simply not enough to ensure commercial viability." Beal stressed however, that he might have remained in business "if the government would have simply guaranteed that NASA's subsidized launch systems would never be allowed to compete with the private sector." Beal views NASA's Space Launch Initiative (SLI) as signaling the death knell for development of new rockets by small, independent companies. The Beal shutdown eliminates the possibility that Beal would launch from Cape Canaveral Air Force Station. Beal's exit from the space launch business is a disappointment for state and local officials which had offered the company a $10 million package of tax breaks and other incentives to build a manufacturing and launch facility at CCAFS. ["Beal Aborts Development of BA-2C Launch Vehicle," *Aviation Week & Space Technology*, October 30, 2000, p 48. "Beal shutdown disappoints," *Florida Today*, November 22, 2000, p 9A-9B.]

◆　A giant lizard is roaming Kennedy Space Center. Karen Foldesi, a Boeing Co. engineer, spotted it while driving on State Road 3. "I spotted what appeared to be a medium-sized alligator crossing the road up ahead, less than a quarter north of KSC's south gate. It was attempting to cross the road on a long curve. There is a pond and some fruit trees on the west side of the road in that area." Southbound traffic was light, she said. "So I slowed almost to a stop to give ample room to this creature stalking across the center line. Driving slowly up closer, I saw it wasn't, it wasn't an alligator at all. When my truck approached it, the large reptile stopped and turned its head in my direction. A long tongue flicked repeatedly out of its mouth." Describing it, she estimates it was about 5 feet long, tail included. "It turned around in the road as I studied it, and then moved rapidly in a blur off the road, through the grass, and crashed into a thicket of Brazilian pepper trees. After looking at dozens of photos on the Internet of Gila monsters and striped and spotted monitor lizards, I believe this is a Komodo Dragon or its close cousin!" The *Encyclopedia Americana* says the Komodo, a native of Indonesia, gets to 10 feet and 300 pounds and often preys upon deer and wild boor — making the KSC/Cape area a hog heaven for it. ["What scaled monster lurks on KSC?," *Florida Today*, October 16, 2000, p 10A.]

◆   The story about Karen Foldesi's seeing what she believed to be a Komodo Dragon on Kennedy Space Center, drew many reactions, including several by other KSC employees who also claimed to have seen what she did. A reserved approach was taken by Mario Busaca, with the NASA Environmental Program Office. He said "We've had some visual sightings of a large lizard of some kind — exactly what we do not know. The sightings are unconfirmed by hard, physical evidence (such as photographs). Given that lack of confirmation, there's not much we can do." ["KSC 'monster' sighted 3 times," *Florida Today*, October 17, 2000, p 10A.]

◆   As soon as he gets approval from the U.S. Fish and Wildlife people at Kennedy Space Center, licensed state trapper James Dean of Melbourne is ready to track down the large and possibly dangerous lizard seen last week along State Road 3 on North Merritt Island. The state has already OK'd it. He is talking with the four people who spotted the powerfully built beast last Thursday and Friday. Three of the sitings were on KSC property. The trapper was told by the wildlife authorities to trap the animal live, because if it is a Komodo Dragon, it is a protected species. ["Trapper has state OK for lizard," *Florida Today*, October 19, 2000, p 12A.]

◆   A former Rockledge businessman has admitted he rigged a bid on a NASA subcontract at Kennedy Space Center seven years ago. Richard E. Finley, the former vice president of R.E.F. Computer Consultants, Inc., pleaded guilty to theft of government property in U.S. District Court in Orlando. Finley was one of three people involved in the conspiracy. Agents from the office of the NASA Inspector General and the Internal Revenue Service uncovered the fraud. The others involved in the conspiracy have already been sentenced. Vernon Baake, a contracting technical monitor with Lockheed Martin, pleaded guilty in January of violating an anti-kickback act. He was sentenced to a year of probation, a $2,000 fine and court costs. Investigators said that in October 1993 Baake gave Finley information about competitors' bids for computer system maintenance for NASA. That allowed Finley to bid slightly lower and win the contract. In return Finley paid a local marina $10,000 for repairs to a boat owned by Baake. In March, Baake voluntarily paid $10,000 restitution to NASA. Ronald Becker of Becker Technology Inc. of Cocoa Beach pleaded guilty in April to submitting false claims of $38,487. Becker was sentenced to two years probation and 180 days of house arrest. He was also fined $2,000, ordered to pay restitution and court costs. Becker entered into an agreement with Finley to do the NASA computer work, submitted false claims on inflated material invoices. Finley, scheduled to be sentenced Dec. 14, could face up to a year in prison. ["Third man pleads guilty to rigging NASA job bids," *Florida Today*, October 14, 2000, p 2B.]

◆   The next president will have a great deal of influence on Brevard County's destiny by helping to shape and finance NASA's plans for future space exploration and determine the course of the commercial space industry in Florida. That, in turn, will have an effect upon the economy, education and other issues. Brevard has seen the economic effects of waiting for new programs to take shape after old ones are discontinued. Neither nominee in this year's presidential election has been pressed on his position on the space industry, but industry officials who track the candidates said they both support the space program. ["Space a key issue to Brevard voters," *Florida Today*, October 7, 2000, p 1A & 4A.]

◆   Russian companies are separately proposing major new systems for the International Space Station. Khrunichev wants to develop a new version of the "functional cargo block" unit, and make it available for a spacecraft that could resupply the station or form the basis of new modules. The latest version of the technology known by its Russian initials, as the FGB, was the first section of the station launched, in a 41-ft.-long, 21-ton module called the Zarya. Another proposal, called the Jig module, is from Energia and Tsniimash. The Jig module would be a mutilpurpose facility that could be used to assemble, launch, service and retrieve smaller spacecraft from the station. It could also support astronauts in extravehicular activity. While the Jig module is in very early stages of development, the FGB may have a better chance of being realized. The effort is predicated on the belief that there will not be sufficient resupply capabilities to the ISS. Currently the shuttle and Progress vehicles are the resupply vehicles for the station. Additional resupply vehicles are being planned by other station participants, including the European Space Agency and Japan's NASDA. The Khrunichev vehicle seems to be a hedge against the failure of the European or Japanese programs. ["Russians propose new space station systems," *Aviation Week & Space Technology*, October 16, 2000, p 59.]

◆   Many of Brevard's space-related businesses will be eligible for significant tax exemptions under legislation recently approved by Governor Jeb Bush. Exemptions are for space flight property leases and for new machinery and equipment used for space technology products and research and development. ["Space industry tax exemption," *Brevard Technical Journal*, October 2000, p 4.]

◆   NASA has awarded four small businesses 90-day contracts totaling $902,000 to develop concepts and requirements to provide access to the International Space Station on emerging launch systems. Established launch services companies are studying the same idea under existing contracts managed by NASA's Kennedy Space Center. ["Small companies to study potential use of emerging launch systems for alternative access to space station," *Brevard Technical Journal*, October 2000, p 4.]

◆   SPACEHAB, Inc., recently embarked upon its 15th Space Shuttle mission with the launch of Atlantis from the Kennedy Space Center. STS-106 was SPACEHAB's third resupply mission to the International Space Station, and its tenth space logistics mission. ["Latest shuttle flight marks SPACEHAB's 15th mission, third space station resupply mission," *Brevard Technical Journal*, October 2000, p 5.]

NOVEMBER

NOVEMBER 1: Workers at Edwards Air Force Base, Calif., began mating the Shuttle Discovery to the Shuttle Carrier Aircraft today and final preparations are being made before departure scheduled for tomorrow morning. Discovery landed at Edwards Air Force Base in California, completing STS-92, the 100th space shuttle mission. ["Discovery won't return before Friday," _Florida Today_, November 2, 2000, p 5A. Bruce Buckingham. (2000). Kennedy Space Center Status Report [Online]. Available E-mail: domo@news.ksc.nsas.gov/subscribe shuttle-status [2000, November 1].]

◆ Space Shuttle Endeavour, which is being prepared for STS-97, is undergoing hot fire testing of its three auxiliary power units. The Rotating Service Structure was extended around the orbiter about 2:00 p.m. today. Tests of the main engines, and helium leaks checks are scheduled for later this week. Orbiter Atlantis is in the OPF high bay 3, being readied for its upcoming STS-98 mission, taking the U.S. Laboratory to the International Space Station. [Bruce Buckingham. (2000). Kennedy Space Center Status Report [Online]. Available E-mail: domo@news.ksc.nsas.gov/subscribe shuttle-status [2000, November 1].]

NOVEMBER 2: A group gathered at Kennedy Space Center on Thursday, November 2, 2000, to dedicate a new high-pressure helium pipeline, which runs from KSC to the Cape Canaveral Air Force Station Complex 37 for Boeing's new Delta IV rocket. SGS will maintain the new 9-mile underground pipeline, which can also serve as a storage tank for shuttle launches. Nearly one launch's worth of helium will be available in the pipeline to support a shuttle pad in an emergency. The line originates at the Helium Facility in KSC and terminates in a meter station at the perimeter of the Delta IV launch pad. ["Boeing Delta 4 launch complex near completion," _Florida Today_, November 3, 2000, p 3A. "Helium pipeline," _Spaceport News_, November 17, 2000, p 2.]

◆ The first U.S./Russian long-duration crew to the International Space Station has embarked on the equivalent of a complex 17-week aircraft flight test program at 235-mi. altitude, following the successful docking of their Soyuz spacecraft with the 80-ton station on November 2. The Expedition One crew, Commander Bill Shepherd, Soyuz Commander Yuri Gidzenko and Flight Engineer Sergei Krikalev, have been busy activating key support systems that supply the station with oxygen, remove carbon dioxide from the air and support communications with the Earth. "Let's go do it, get those shuttles ready," Shepherd said on launch day as he climbed the pad stairs to enter the Soyuz TM-31 vehicle. Three shuttle/ISS assembly missions are to visit the station during Expedition 1's four-month stay, two of which will increase dramatically the ISS' capability. The station on which the Expedition 1 crew arrived will be vastly different when they leave in February. The first four weeks aloft are designated "Stage 2R" and will be the primary phase to turn the ISS into a habitable spacecraft, as opposed to one capable of being visited temporarily as it has been on the last five shuttle flights. Much of the first crew's mission is to get the station ready for future crews. At the moment the ISS crew's Soyuz rocket was lifting off, top Russian aerospace industry and government officials were meeting to chart more consolidation and cutbacks in the bloated Russian aerospace infrastructure. Expedition 1 marks only the second time that a U.S. astronaut has been launched on a Soyuz rocket. The first was Norm Thagard's mission to Mir in 1995. The orbiter Endeavour, set for a Nov. 30 liftoff, STS-97, was rolled to its Kennedy Space Center pad the same day that Expedition 1 was launched. Endeavour's STS-97 crew will install a set of 240-ft. solar arrays, and STS-98 will deliver the school-bus sized Boeing laboratory module. The Expedition 1 crew will return to Earth on Discovery's STS-102 mission. ["ISS Finally Manned as Challenges Abound," *Aviation Week & Space Technology*, November 6, 2000, p 30-32. "ISS Crew Must Merge U.S., Russian Systems," *Aviation Week & Space Technology*, November 6, 2000, p 30-31. "Crew set to arrive at station today," *Florida Today*, November 2, 2000, p 1A & 3A. "NASA Updates Expedition One Progress," *NASA News Release #N00-056*, November 7, 2000.]

◆   Orbiter Discovery, mounted atop NASA's Shuttle Carrier Aircraft departed Edwards Air Force Base, Calif., today at 11:51 a.m. EST. The ferry flight made a refueling stop at Altus AFB, Oklahoma, at 2:56 p.m. EST today. The afternoon weather briefing indicated the possibility of unacceptable weather in the Altus AFB vicinity, so ferry managers decided to overnight at Whiteman AFB, Missouri. Ferry flight rules state the orbiter and SCA cannot fly through precipitation, thick clouds or high turbulence. There are also wind and temperature restrictions. [Bruce Buckingham. (2000). Kennedy Space Center Status Report [Online]. Available E-mail: domo@news.ksc.nasa.gov/subscribe shuttle-status [2000, November 2].]

◆   On Kennedy Space Center's Launch Pad 39B, engineers will conduct main engine flight readiness test on Shuttle Endeavour, in preparation for STS-97. [Bruce Buckingham. (2000). Kennedy Space Center Status Report [Online]. Available E-mail: domo@news.ksc.nasa.gov/subscribe shuttle status shuttle-status [2000, November 2].]

◆   During its first free flight, NASA's unmanned X-38 vehicle 131R did a slow 360-degree roll after release from its B-52 carrier aircraft on November 2. It was automatically stabilized by the preprogrammed deployment of a drogue chute and made a successful landing under parafoil on a dry lakebed runway at Edwards AFB, Calif. The vehicle which sustained no damage, is an 80% scale version of the CRV designed to provide emergency escape for International Space Station crews. ["World news roundup," *Aviation Week & Space Technology*, November 6, 2000, p 24.]

NOVEMBER 3: Shuttle Discovery spent the night at Whiteman AFB in Missouri before continuing its journey to Kennedy Space Center. Discovery left Edwards AFB, Calif., at 11:51 am EST Thursday, November 2, on a modified Boeing 747, and made a refueling stop at Altus AFB in Oklahoma, three hours later. Due to bad weather, managers decided to fly east to Missouri, and resume the flight Friday morning. Pilot Bill Brockett guided the pair to a smooth landing at KSC at 4:35 p.m. Ground crews are preparing to demate Discovery from the SCA tonight and plan to transfer the orbiter to Orbiter Processing Facility bay 1 tomorrow. Discovery and the SCA are currently located in the mate-demate device at KSC's Shuttle Landing Facility. On October 24, Discovery ended its 13-day STS-92 mission to the International Space Station in California, due to bad weather at KSC. ["Discovery headed home from California," *Florida Today*, November 3, 2000, p 3A. "Discovery returns to KSC," *Florida Today*, November 4, 2000, p 3A. Bruce Buckingham. (2000). Kennedy Space Center Status Report [Online]. Available E-mail: domo@news.ksc.nasa.gov/subscribe shuttle-status [2000, November 3].]

NOVEMBER 4: Kennedy Space Center officials estimate nearly 44,000 people came to KSC and the Cape Canaveral Air Force Station on Saturday, November 4, to get a close-up look at shuttle Endeavour on the launch pad, take a peek inside the Vehicle Assembly Building, and shake hands with an astronaut. The joint Community Appreciation Day was a phenomenal success. The biggest draws of the day were Launch Complex 39B, where Endeavour was being prepared for its November 30 launch, and briefings from seven astronauts who spoke at various sites around the center. ["KSC open house lures nearly 44,000 visitors," _Florida Today_, November 5, 2000, p 1B.]

NOVEMBER 6: The crew for mission STS-97 arrived at KSC to begin the Terminal Countdown Demonstration Test (TCDT). They will be going through emergency egress training, familiarization with the payload, and a simulated launch countdown, before returning to Houston on Wednesday afternoon. Over the weekend, Endeavour's three auxiliary power units were successfully hot fired. Solid rocket booster frequency response test activities are complete as well. The Shuttle's helium signature leak test is ongoing, and preparations to load the orbiter's onboard hypergolic storage tanks are under way. ["STS-97 crew arrives at KSC for TCDT," _KSC Countdown_, November 7, 2000. "Shuttle crew arrives at KSC," _Florida Today_, November 7, 2000, p 1B. Bruce Buckingham. (2000). Kennedy Space Center Status Report [Online]. Available E-mail: domo@news.ksc.nsas.gov/subscribe shuttle-status [2000, November 6].]

◆   Following a successful landing of the Shuttle Carrier Aircraft on Friday afternoon, Orbiter Discovery was demated from the modified Boeing 747 and towed to the Orbiter Processing Facility Saturday afternoon. The orbiter's payload bay doors are open and workers are removing the Imax 3-D camera so that it can be turned around for use on STS-97. Today, the orbiter's tail cone is being removed and power system validation is scheduled to occur. [Bruce Buckingham. (2000). Kennedy Space Center Status Report [Online]. Available E-mail: domo@news.ksc.nsas.gov/subscribe shuttle-status [2000, November 6].]

NOVEMBER 7: The next astronauts to visit space station Alpha say they are not concerned with safety at Kennedy Space Center despite the recent discovery of a pin found on shuttle Discovery before its last launch. Echoing their predecessors, the crew of shuttle Endeavour, set to launch November 30, say that certain hazards come with the job. The United Space Alliance is heading an investigation to find out how the four-inch metal pin fell on the ET fuel line. ["Astronauts ready for any situation," _Florida Today_, November 8, 2000, p 1B.]

NOVEMBER 8: A ribbon-cutting on Wednesday, November 8, officially reopened the site for the ELV Program Office at the newly renovated E&O Building. The Expendable Launch Vehicle Program was transitioned to KSC in October 1998. The E&O Building was home for NASA's unmanned missions since 1964, and renovations to the building began in August 1999 to correct aging infrastructure problems and to make the building handicapped accessible. The ELV Launch Services moved back into the renovated E&O in October 14, 2000. ["ELV Program Office Gets a New Home," _KSC Countdown_, November 7, 2000. "E&O ribbon cutting," _Spaceport News_, November 17, 2000, p 2.]

◆   Kennedy Space Center Director Roy Bridges and U.S. Air Force 45th Space Wing Commander Brig. Gen. Donald Pettit signed the Consolidated Comprehensive Emergency

Management Plan (CCEMP) on November 8. The CCEMP established uniform policy guidelines for the effective mitigation of, preparation for, response to and recovery from a variety of emergency situations at the Cape Canaveral Spaceport. ["Emergency Plan signed," *Spaceport News*, November 17, 2000, p 2.]

◆　At Launch Pad 39B, the Terminal Countdown Demonstration Test with Endeavour's flight crew of STS-97 was concluded. Leak checks on Atlantis' three main engines are now in work, and Discovery's tail cone has been removed. [Bruce Buckingham. (2000). Kennedy Space Center Status Report [Online]. Available E-mail: domo@news.ksc.nsas.gov/subscribe shuttle-status [2000, November 8.]

NOVEMBER 9: The fourth annual Space Coast Birding and Wildlife Festival offered tours and seminars for nature lovers on a variety of topics. From Thursday, November 9th through Sunday, November 12th, seminars about and tours of undeveloped areas of Kennedy Space Center, the Merritt Island National Wildlife Refuge, and other nearby nature preserves were offered by four KSC employees among others. ["Wildlife fest offers tours, seminars," *Florida Today*, November 5, 2000, p 1B.]

◆　The Boeing Company of Seattle, WA, and United Space Alliance ("USA") of Houston, TX, agreed to pay $825,000 and to give up their rights to $1.2 million in unpaid invoices to settle allegations relating to false claims submitted to NASA between 1986 and 1992. The settlement is based on a civil lawsuit filed by the U.S. Department of Justice on January 11, 2000, in the Central District of California. The suit alleged that Rockwell Space Operations Company ("RSOC") knowingly passed along the fraudulent costs of its subcontractor, Omniplan Corporation, under the NASA Space Shuttle and Space Station Freedom Programs. Between 1986 and 1993, RSOC certified to NASA that all Omniplan costs were reasonable, allowable, and allocable to the space agency contracts. The United States alleged that the RSOC invoices included large amounts of fraudulent costs, including personal homes, a ski lodge, expensive jewelry, and vacation expenses. Because Boeing and USA are the successors of RSOC, both contractors have now assumed legal responsibility for the fraudulent costs passed along to NASA by RSOC. ["NASA Contractors Agree to Settlement Values at $2 Million," *NASA News Release #2001-013*, November 9, 2000.]

NOVEMBER 10: A report finalized Friday, November 10, concluded that the pit pin which delayed last month's launch of shuttle Discovery's STS-92 mission, came from the Vehicle Assembly Building. A pin which helped to secure a platform above part of the shuttle's fuel tank was installed incorrectly. When the platform was moved, September 8 to allow the shuttle to leave the VAB for the launch pad, the four-inch metal pin apparently fell 73 feet and landed on a fuel line near one of the tank's connections to the shuttle. The pin stayed there until it was found by inspectors October 10. Fearing the pin could fall and damage the shuttle during launch, NASA delayed Discovery's liftoff one day so the pin could be removed. The investigation team recommended a more regimented inspection of the shuttle to look for objects that shouldn't be there, such as the pin. Before shuttles leave the VAB for a launch pad, inspection teams will do a meticulous top-to-bottom check for foreign objects. Another recommendation was to make sure all of the pins work properly. There are thousands of such pins at KSC, and about 200 have been repaired since the errant pin was found at the launch pad. ["Pin found on shuttle from VAB, report says," *Florida Today*, November 11, 2000, p 1B.]

◆   The five-man crew of shuttle Endeavour found out it will have another spacewalk added to its STS-97 mission. The added spacewalk, which will extend the flight by another day, will attach an experiment measuring plasma particles around the International Space Station and a set of new solar arrays. ["Spacewalk added to shuttle mission," *Florida Today*, November 14, 2000, p 1B.]

NOVEMBER 13: The P6 integrated truss segment, payload on mission STS-97, was transferred to the payload transport canister, and will be moved to the launch pad on November 14. The payload will be installed in Discovery on November 16. During two EVAs the crew will install the P6, which includes the first two U.S. solar arrays and a power distribution system, plus two radiators that will provide early cooling on the ISS. ["STS-97 payload moves to Launch Pad 39B," *KSC Countdown*, November 14, 2000.]

NOVEMBER 14: Six years ago, acknowledging that commercial space launches are increasing and that companies could look to other countries to send their satellites into orbit if Florida doesn't update its launch facilities, officials put together the Cape Canaveral Spaceport Symposium. At this week's sixth annual symposium, industry executives, state and federal officials, and Air Force and NASA representatives, will meet to find the best way to make Cape Canaveral a spaceport for the future. The machinery and electronics that track rocket launches from the Cape is stretched to its capacity by the current launch schedule, and government paperwork is stretching launch companies' patience. Discussions will include ways to ease bureaucratic burdens of launching from the Cape, updating outdated tracking equipment to allow more launches, and how to pay for modernization. ["Session's goal: keep launches in Florida," *Florida Today*, November 13, 2000, p 1B. "Cape could be spaceport of the future," *Florida Today*, November 15, 2000, p 1B.]

◆   In preparations for the STS-97 mission, the P6 payload arrived at Launch Pad 39B today and is being installed in the payload changeout room. [Bruce Buckingham. (2000). Kennedy Space Center Status Report [Online]. Available E-mail: domo@news.ksc.nsas.gov/subscribe shuttle-status [2000, November 14].]

NOVEMBER 16: Russia's Cabinet decided that space station Mir will end its 15 years in space with a fiery plunge into the Pacific Ocean in February. For years NASA has been urging Russia to concentrate its funds on the new International Space Station. Russian Space Agency chief Yuri Koptev said an unmanned cargo ship would be sent to Mir in January, and in February the cargo ship would fire rockets to push Mir into the atmosphere. ["Russia plans to ditch Mir off Australia," *The Orlando Sentinel*, November 17, 2000, p A-5.]

◆   The STS-97 mission's P6 truss and solar arrays is being installed in Endeavour's payload bay today. ["STS-97 payload being installed in Endeavour," *KSC Countdown*, November 16, 2000.]

◆   The Florida Board of Regents has created the National Alliance for Aerospace Science and Education, giving strong state blessing to a partnership among the University of Central Florida (UCF), the University of Florida (UF). and NASA to create a broad research and education center at Kennedy Space Center. The designation allows the alliance to apply for state money and comes with an initial request for $1 million in seed money from the

Legislature. Officials say that should quickly help attract millions more from NASA and other federal agencies. UCF's portion of the space partnership will be built around its five-year-old Florida Space Institute, which focuses on launch and space-travel research and education. UF brings its Space Education and Research Laboratory into the picture. UF plans to build a $28 million lab at KSC, which would focus on biological research in space. Together, UCF and UF will coordinate research efforts from any Florida college or university that wants to have a presence at KSC. The new partnership should help a NASA policy to turn KSC into a research center. ["UCF is seeing stars in its future," *The Orlando Sentinel*, November 17, 2000, p B-1 & B-4.]

◆   A group of 29 Clay High School students from South Bend, Ind., was transported to the emergency room at Parrish Medical Center, in Titusville after becoming ill while visiting Kennedy Space Center. The students were on a Reserve Officers Training Corps field trip. ["29 teens ill at space center," *The Orlando Sentinel*, November 17, 2000, p B-2.]

NOVEMBER 18: More than 100 Boy Scouts from Florida, Alabama, and Georgia attended Kennedy Space Center's Overnight Adventure program Saturday and Sunday. The program is set up to give the Boy Scouts and other groups a close look at the space program by offering them a spectacular after-hours crash course in space exploration. Astronaut Vance Brand talked with the scouts and joined in the overnight camp out. ["Boy Scouts sleep over at KSC," *Florida Today*, November 20, 2000, p 1B & 2B.]

◆   An automatic docking of a 7.5-ton Progress resupply vehicle to the International Space Station was aborted. The ISS crew took emergency actions to get the crucial resupply vehicle docked to the station on Nov 18. The ISS crew received high praise for the way it handled the situation. The Progress was launched from Baikonur Cosmodrome on Nov. 16, with 2 tons of logistics as part of the initial activation of the station by the Expedition 1 crew. The Progress had to dock with a different port than usual, but when the docking program shifted to the short-range phase, the transport began oscillating, apparently carrying out search patterns to acquire final guidance data. Flight controllers told Yuri Gidzenko to take over manual control of the docking. Using visual observations by other ISS crew members, because his video systems were obscured by water vapor and bad lighting angles, Gidzenko brought the Progress within 8-10 meters of the ISS. The crew was then told by flight controllers to continue in a station-keeping mode until lighting improved. Gidzenko was then able to bring the Progress in for a successful docking. Victor Blagov, deputy flight director, praised Gidzenko and the ISS crew. "I would like to congratulate you for your bravery and heroism, and your ability to be patient." ["ISS Crew Saves Progress, Readies Station for Shuttle," *Aviation Week & Space Technology*, November 27, 2000, p 40 & 45.]

NOVEMBER 24: Space station Alpha crew members finished unloading 2 tons of supplies from a Progress supply ship. They will refill the craft will trash and jettison it to burn up in the Earth's atmosphere. The unloading has been rushed because the Progress ship would interfere with docking the Endeavour on its STS-97 mission to the station. The crew must now stow the supplies before the scheduled docking with Endeavour. ["Alpha stocked with supplies," *Florida Today*, November 25, 2000, p 1B.]

◆   Brian Welch, a longtime NASA public affairs officer, died of a heart attack Friday, Nov. 24. As Director of Media Services, Welch, 42, led many of the agency's public outreach efforts. Welch was also one of the voices of Mission Control in Houston during earlier shuttle missions. He became Director of Media Services in 1998. ["NASA public affairs officer suffers fatal heart attack," *Florida Today*, November 28, 2000, p 1B. "Brian Welch, NASA Director of Media Services, Dies," *NASA News Release #00-187*, November 27, 2000.]

NOVEMBER 26: NASA sponsors programs such as the Summer High School Apprenticeship Research Program, robotics competitions, and an Internet surfing contest. Some teens think it will take more than science to attract the typical teen, but it seems unlikely that new attractions at the Kennedy Space Center Visitor Complex will have a Disney fantasy appeal. "Our job is to tell the NASA story," explained Dan LeBlanc, director of marketing for the Kennedy Space Center Visitor Complex. "It has to be based on science." LeBlanc says KSC is developing a formal education program directed at teens, and several new attractions which should appeal to teens. ["Future projects focus on science not fantasy," *Florida Today*, November 26, 2000, p 12D.]

◆   Shuttle Endeavour is scheduled to lift off at 10:06 p.m. on Thursday, November 30, carrying enormous winglike solar arrays which will convert the sun's rays into electricity to power the International Space Station. The 17-ton, 49-foot-long array will be one of the heavier payloads the shuttle has carried into space. Because of the payload's weight, NASA can only send five astronauts, rather than the usual seven, into space. ["Shuttle's crew will have a busy mission," *Florida Today*, November 26, 2000 p 1A & 8A.]

NOVEMBER 27: Endeavour's crew members were surprised that reporters and photographers were on hand when they arrived at Kennedy Space Center on November 27, given the presidential drama unfolding in Florida. "We appreciate you all coming out," Commander Brett Jett said. "We know there's another big story going on in Florida today." Crew members arrived about 30 minutes late, because their T-38 jets had mechanical problems prior to takeoff from Ellington Field in Houston. ["Endeavour's crew ready for Thursday launch," *Florida Today*, November 28, 2000, p 1A.]

NOVEMBER 28: A hectic launch schedule has NASA officials taking extra precautions to prevent delays or mishaps, as the agency prepares for its third launch in as many months. Agency officials cautioned Tuesday that while they're confident they can maintain the pace, many things could go wrong to delay Shuttle Endeavour's Thursday night liftoff. NASA is in the midst of a nine-launch stretch, using four orbiters in only 11 months to build the International Space Station and conduct other experiments. Some worry the torrid pace could lead to human errors or accidents, as NASA faces its toughest schedule in three years. The launch of STS-92 last month was delayed for a day after a pin was noticed on Discovery's fuel tank. As a result, new steps have been implemented to make sure that does not occur again. Shuttle Program Manager Ron Dittemore said "After a problem like that...there probably isn't a loose pin from here to Orlando". ["Busy launch schedule keeps KSC on its toes," *Florida Today*, November 29, 2000, p 1A & 2A.]

NOVEMBER 30: Endeavour and the crew of STS-97 was launched from Kennedy Space Center's Launch Pad 39B at 10:06 p.m. Endeavour will be commanded by Brent Jett (Cdr., USN) and piloted by Mike Broomfield (Lt. Col., USAF). Mission specialists include Joe Tanner, Carlos Noriega (Lt. Col., USMC) and Marc Garneau from the Canadian Space

Agency. The mission to the International Space Station will deliver and install massive solar arrays which will increase the amount of electricity for the station by five-fold. Covered by 65,000 silicon cells, the arrays will require two spacewalks by the shuttle's crew to install. Installing the array is a critical step in the $95 billion, 16-nation effort to build a laboratory complex in space by 2006. A third spacewalk to attach an experimental pod to the top of the solar tower is also planned. Endeavour is scheduled to return to KSC on December 11. As usual, future shuttle missions to build the station depends on how this one goes. Crew members are confident they are more than ready to get the job done. ["Shuttle go for liftoff tonight," _Florida Today_, November 30, 2000, p 1A & 8A. "Shuttle crew to boost Alpha power," _The Orlando Sentinel_, November 30, 2000, p A-1 & A-14.]

◆   Shuttle Endeavour's 17-ton cargo would make an exceptionally dangerous emergency landing even more difficult. If something goes wrong during the early phases of the launch tonight, and the shuttle has to land, the added weight of the heavy payload, would make landing more difficult. Extra weight would change the center of gravity, making the shuttle fly differently, and putting more stress on the wings. NASA has successfully flown heavier payloads into space, with the record-holder being the STS-93 mission carrying the 50,000-pound Chandra X-ray telescope. If the astronauts are unable to install the solar arrays on the station, it would have to be repacked in the cargo bay and returned to Earth. There have been successful re-entries with heavy cargoes also. ["Cargo adds to emergency risk," _Florida Today_, November 30, 2000, p 1A.]

◆   NASA has placed tiny cameras on the helmets of the two astronauts scheduled for spacewalks on STS-97. The cameras will provide bird's-eye views of what astronauts Carlos Noriega and Joe Tanner are doing. Earthbound people will see hands working with power tools, and what the astronauts see while they work with delicate pieces of hardware in the weightlessness of space. The cameras will also help engineers help the astronauts when unexpected problems require solutions. ["Cameras to show spacewalker's view," _Florida Today_, November 30, 2000, p 8A.]

DURING NOVEMBER: Early next year an all-weather spaceport will become operational in Alaska's Kodiak Island, offering commercial competition to government facilities at Cape Canaveral, Fla., and Vandenburg AFB, Calif. The Kodiak Launch Center is an ideal location for launching low-Earth orbit communication satellites, which will offer customers lower prices and a higher degree of scheduling flexibility. Payloads of up to 8,000 lb. can be launched into polar and highly elliptical orbits. The biggest advantage of the site is its unobstructed downrange launch corridor. Launches are to the south with no populated areas for thousands of miles, which gives the range safety officer longer to observe a rocket to see if it can be saved, before having to destroy it. The complex was designed for all-weather capability, so all prelaunch preparation and processing for the payloads and rocket motors is done inside. The launch vehicles and payloads will not be exposed to the elements until just prior to launch. The complex was developed on a very low budget by the Alaska Aerospace Development Corp., a public entity created with state support in 1991. It has four launches scheduled for 2001. Commercial customers choosing Kodiak will not have to compete with the Defense Dept. for launch times as they would at Vandenberg or Cape Canaveral. ["Alaska Competes for Satellite Launches," _Aviation Week & Space Technology_, November 13, 2000, p 77 & 79.]

◆　School children in Pike County, Ala., and Memphis, Tenn., are getting more computers in their classrooms thanks to a recent distribution of 690 Pentium I computer systems from the NASA KSC Property Disposal Office. The unusually large number of computers became available because of computer changeouts made by United Space Alliance (USA) and ODIN. Computers are donated to schools through the federal Computers for Learning program. ["Computer donation benefits schools," *Spaceport News*, November 17, 2000, p 2.]

◆　NASA has selected Command and Control Technologies Corporation (CCT), Titusville, to conduct research on integration of space launch vehicles with the worldwide air traffic control system, pending successful award negotiations. The "phase II" Small Business Innovative Research (SBIR) project will concentrate on developing methods for managing the flight of reusable launch vehicle operations through commercial airspace. Embry-Riddle Aeronautical University will provide development and consulting services related to air traffic management techniques. ["CCT selected for space launch range research," *The Brevard Technical Journal*, November, 2000, p 18-19.]

◆　Florida Tech's Fall Humanities Lecture Series recently presented "NASA History on the Lighter Side: Political Cartoons and American Culture." Dr. Roger Launius, NASA's chief historian, presented a history of human space flight using political cartoons drawn from throughout the history of the space age. ["Florida Tech lecture series explores history of NASA," *The Brevard Technical Journal*, November, 2000, p 26.]

DECEMBER

DECEMBER 2: The planned launch of an Atlas 2AS rocket, carrying a National Reconnaissance Office satellite, has been delayed at least a day because of hardware issues. An engine tested at a factory but similar to the one used on this mission had a problem so follow up tests are being done to the rocket on the pad. Launch, initially planned for December 4, has been pushed back to December 5. ["Problem delays Atlas launch," *Florida Today*, December 3, 2000, p 3A.]

DECEMBER 4: STS-98 shuttle Atlantis was rolled over to the Vehicle Assembly Building and lifted into high bay 3 for stacking with the external tank and solid rocket boosters. It is expected to start rollout to launch pad 39A December 11 at 7 a.m. That same day, Endeavour (STS-97) is scheduled to land at the Shuttle Landing Facility at 6:19 p.m. after its 11-day mission to the International Space Station. ["STS-98 mission in sight; Atlantis mated to stack in VAB, to roll out Dec. 11," *KSC Countdown*, December 5, 2000.]

DECEMBER 5: An Atlas 2AS rocket successfully launched from pad 36A at Cape Canaveral Air Force Station at 9:47 p.m. The scheduled time was between 8:14 and 10:12 p.m. The launch was delayed several times due to a boat in the launch hazard area, a failed heater on stabilizing gyroscopes, a problem with an air conditioner and computers that had to be adjusted for higher than expected upper-level winds. The payload was a satellite for the National Reconnaissance Office. ["Atlas 2 lifts off with spy satellite," *Florida Today*, December 6, 2000, p 1B.]

DECEMBER 7: During the STS-97 launch, one of four explosive bolts holding the left solid rocket booster (SRB) to the external tank did not separate on first command. A backup explosive on the same bolt did fire safely separating the rocket. The defective bolt was discovered during a routine inspection of the SRB. Flight Director Bill Reeves said similar explosives have been used to separate rocket stages since the Gemini Program began in 1965. ["Faulty booster bolt uncovered," *Florida Today*, December 9, 2000, p 1A.]

◆ The S3 Integrated Truss Structure (ITS), the second starboard truss segment, has arrived at KSC's Shuttle Landing Facility aboard a Super Guppy aircraft. It was transferred to the Operations and Checkout (O&C) Building. The S3 truss is scheduled to be added to the Space Station in April 2003. ["New Space Station element arrives at KSC," *KSC Countdown*, December 12, 2000.]

DECEMBER 10: Kennedy Space Center's security SWAT team took 11th place in the 18th Annual SWAT Roundup held in Orlando last week. Eighty-three teams from around the world matched skills in five different events. ["SWAT teams meet to learn, compete," *Florida Today*, December 10, 2000, p 1B.]

DECEMBER 11: After successfully completing all objectives of the STS-97 mission, shuttle Endeavour touched down at 6:04 p.m. EST on KSC's Shuttle Landing Facility Runway 15. Coincidentally, space station Alpha passed over KSC just as the sonic booms from Endeavour reached the ground. This was the 16th nighttime landing for the Space Shuttle and the 53rd at Kennedy Space Center. ["Mission complete, Endeavour and crew